The Evolution of Culture in Animals

John Tyler Bonner

Original drawings by Margaret La Farge

PRINCETON UNIVERSITY PRESS

PRINCETON, NEW JERSEY

Preface

The stimulus for this book came in 1975 when I was invited to a Daedalus Conference organized by Talcott Parsons and Hunter Dupree on the relation of biological evolutionary theory to social science. From this small meeting of anthropologists, sociologists, psychologists, neurobiologists, and evolutionary biologists I gained a full appreciation of my own confusion on the subject and a desire to do something about it.

The first stage was writing a short essay that was never published, but that produced numerous comments and criticisms both from social scientists and biologists who were kind enough to read it. In this groping fashion I began to see how things fit together, at least in my own mind, and I half decided to work on the problem further, possibly in the form of a book. What decided me finally to take the plunge was the possibility of a leave from Princeton University and the encouragement of my good friend George Berry.

I began the first draft in our small house in Cape Breton and finished it at the University of Edinburgh in the stimulating environment of the zoology department. There I received endless help and advice from my many friends. In particular I want to mention John Godfrey who time and again drew my attention to some key bit of information that always seemed to be just what I was seeking. But I must also thank many others, especially Phillip Ashmole, Aubrey Manning, and Linda Partridge for many helpful discussions. My greatest debt is to Murdoch Mitchison who made my stay possible in his department, and who, with many kindnesses, saw to it that it was maximally pleasant and profitable. The School of Epistemics at the University of Edinburgh asked me to give a general lecture on the subject of my book that was very helpful to me in learning how to communicate my thoughts. I had the good sense to give the lecture the day before I left town to return to Princeton.

This preliminary draft was circulated either *in toto* or in parts to a large number of people who sent me criticisms. Furthermore, in the spring of 1978 a group of social scientists at Princeton held a weekly seminar on sociobiology and I spent three of those sessions discussing my draft. There were so many people involved in that public-spirited enterprise that it is impossible to list them all, but I owe each one of them a large share of gratitude. I should particularly like to mention James Beniger, John Burns, Roberta Cohen, Peter Huber, Susanne Keller, Marion Levy, Norman Ryder, Susan Watkins, Maxine Weiss, Robert Wuthnow, and all the others who contributed to the meetings. Also I want to give special thanks to my old friend Marion Levy for spending so much time with the manuscript and giving me a better understanding of how I could say something that might be useful to the sociologist. I am also grateful to Hunter Dupree and Talcott Parsons, not only for starting me on this course, but giving me good comments on the draft.

Among my biological colleagues my list of critics is even greater. In particular I want to thank Henry Horn, Aubrey Manning, E. O. Wilson, and an anonymous reader. Also many helped me with specific points: Joan Aaron, Edward Cohen, John Endler, Alan Gelperin, James Gould, Robert May, W. G. Quinn, and Thomas Sanders, among others. I would like to single out Henry Horn's major contribution, for which I am truly grateful. His marginal notes were worthy of being set into print intact had not some of them been so uncomplimentary.

And I have not exhausted the list. Edward Tenner of Princeton University Press gave me numerous good ideas and advice. Of particular help was my wife Ruth's meticulous attention to inconsistencies in my grammar, style, and thought, and she was unfailingly diplomatic in giving me the bad news. I also thank Kathryn Schneider for her library research, her most helpful comments on the first draft, and the skilled preparation of the first two drafts. It has been a special pleasure working with Margaret La Farge who did all the splendid original drawings for the book.

I have never before been helped by so many kind and forebearing people, but I do not want to imply that the final draft of the book was written by a committee. While in many places I have been saved by my friends from making absurd and embarrassing statements, in others I retained the right to do so, as the reader may enjoy discovering. The statement in this book is a personal document. It reflects what seem to me, as a biologist, the important issues that lie between the biological and the social sciences. It avoids political issues and avoids all those issues that have produced such intense emotion in the last few years. This is not because I don't take a certain morbid fascination in such popular matters; it is only that it is not possible to discuss them constructively at this time. For the moment they have removed themselves from the realm of science, and this book is about science.

Finally, and I have saved it to the last because of its importance, let me acknowledge with gratitude the Josiah Macy, Jr. Foundation for its grant supporting the preparation of this book. The president, Dr. John Z. Bowers, was unfailing in his enthusiasm for the project, and this did much to keep my spirits up when the going seemed slow.

Margaree Harbour
Cape Breton
Nova Scotia

J.T.B.

The Evolution of Culture in Animals

CHAPTER 1

Philosophy and Less Grand Matters

This is a book on evolutionary biology. It stresses certain aspects in a way that will perhaps provide some new insights into old problems. I say "perhaps" because there are many sophisticated evolutionary biologists who will point out that there are really no new or revolutionary ideas; what I have to say is essentially what they knew all along. That is most likely the case; yet they have not put it all quite this way. The difference will be in the grouping and arrangement of facts and ideas. It is often true that, as a subject advances (especially when it advances rapidly), we do not always appreciate immediately all the riches we have before us.

While it is my hope that I can convince the professional biologist that there are new things to learn from old facts, I also want to reach the anthropologist, the sociologist, and the well-read layman. Therefore I will couch my arguments in as little jargon as possible. This is not to mean that I subscribe to vagueness or that I court imprecision. To the contrary, I want to make my exact meaning immediately clear.

My purpose is to trace the origins of the human cultural capacity back into early biological evolution. It will soon be evident that I am not a catastrophist and do not believe that, like the great flood, culture suddenly appeared out of the blue at a restricted moment in the early history of man. Rather, to continue to borrow from the nineteenth-century geologists, I am a uniformitarian and believe that all evolutionary changes were relatively gradual and that we can find the seeds of human culture in very early biological evolution.

A Brief Abstract of the Book

There is a main point, a principal conclusion to this book, and I should like to state it in the beginning so that it will be easier to follow the thread of my argument. It is that even though culture itself does not involve genetic inheritance or, therefore, Darwinian evolution by natural selection, the ability of any animal to have culture is a direct product of such an evolutionary mechanism. Passing information by behavioral rather than genetic means has made it possible in some cases to pass kinds of information that either cannot be transmitted genetically at all or are less effectively transmitted by genes. Natural selection operates on the genes and only involves gene transmission; yet the power to transmit by behavioral means is as a method adaptively advantageous. Therefore, there has arisen a genetically determined behavioral capacity to transmit information by signs, by language, by imitation.

This kind of nongenetic transmission is mediated by the brain, and so there has been a selection pressure for a larger and more complex brain. The advantage of culture both in its present form as seen in man, and in its more primitive forms as seen in other animals, has continuously exerted pressure for brain expansion. As a result there was first an increase in the ability to learn and later in the ability to teach. A related trend can be seen in the increase of the flexibility of responses ranging from single, rigid, genetically determined responses to multiple responses that can either be learned or even invented. These trends form a progression ultimately leading to culture. I shall also examine the early origins of these trends and follow the complete sequence of biological events that are the antecedents of culture.

Culture is transmitted by behavioral rather than genetic means, and it will be important to keep this distinction clear. The problem is that any pattern of behavior could have both a genetic and a learned or acquired component, which involves the old intractable question of nature versus nurture. Many reasons can be given for the intractability, but none of them makes the question any less interesting, and human nature is such that we shall continue to try and find ways of identifying what is inherited and what is learned. However, for various reasons, involving both its difficulty of analysis and the intellectually destructive political emotions it generates, this is a subject I shall be careful to avoid in this book. Instead I shall ask why culture exists at all, a question that can be answered in the straightforward Darwinian manner just indicated.

Reductionism and Holism

The conflict that has arisen between biology and the social sciences can, in large measure, be seen in terms of the conflict between reductionism and holism. By reductionism we mean a science (or a hierarchical level) can be understood in terms of its component parts from the level below; for instance, the symmetrical structure of a crystal can be interpreted largely from the properties of its constituent molecules. By holism we mean that there are emergent properties arising at each hierarchical level and that these properties cannot be understood in terms of those of a lower level. The holist believes the living organism has properties that would not be predictable on the basis of what we know of chemical substances and the characteristics of human society cannot be interpreted in terms of lower level biological properties. The old adage of holism is that the whole is more than the sum of its parts.

Biology has undergone, in its most recent flowering, a period of extraordinarily successful reductionism. It is hard to know exactly when this began, but let me give a few milestones: the interpretation of heredity in terms of unit characters by Gregor Mendel; the demonstration by T. Boveri, E. B. Wilson, W. S. Sutton, T. H. Morgan, and others that those characters resided in the chromosomes of the nucleus; the discovery of the structure of DNA by J. Watson and F.H.C. Crick; and finally the cracking of the genetic code by M. W. Nirenberg and others who showed which triplets of nucleotides in the DNA specified particular amino acids in the proteins. By any measure these and other advances in molecular biology have been staggering and at this very moment the rapid progress continues unchecked.

If we turn to evolutionary biology, there has been a similar trend, although it is less spectacular in its progress. Its origins of course can be traced to Charles Darwin. The next step forward was the rise of population genetics in the 1920s and 30s, especially the work of R. A. Fisher, J.B.S. Haldane, and Sewall Wright. This era of neo-Darwinism was truly reductionist because its concern was for the rates of change of individual genes in a population over time. In the 1940s and 50s the field was criticized because it was oversimplified; it did not seem to reflect the real world, and therefore was unable to cope with what were perceived as the new and more significant problems.

The next surge forward came in the 1960s when Robert MacArthur and his colleagues and followers saw that one could make simple theoretical models that applied not only to the more complex aspects of evolution, but in particular to the morass of problems in ecology. Their method of simplification and approximation dramatically illuminated possible mechanisms in a way that made order out of chaos. There was great resistance from the traditional evolutionists and ecologists because it seemed that the very complexity of the problems, which had been cherished as their most important characteristic, was consciously ignored. MacArthur (1968) countered these arguments with: "Think how physics would be without its frictionless pulleys, conservative fields, ideal gases" (1968:162). Theoretical ecology and evolution have already proven, in a very short time span, to be enormously powerful as analytic tools.

Nevertheless its proponents have not rejected holism, but they believe that progress can most effectively be achieved by a balance between holism and reductionism. Again to quote MacArthur (1972): "Most scientists believe that the properties of the whole are a consequence of the behavior and interaction of the components. This is not to say that the way to understand the whole is always to begin with the parts. We may reveal patterns in the whole that are not evident at all in its parts. Species diversity, for example, is a community property and is not a property of the individual component species. It can be understood as a consequence of the interaction of these species, but its patterns were discovered and explained by people aware of communities; ecologists primarily interested in the separate species have never made any progress in unravelling community patterns" (1972: 154, 155). In his own research MacArthur showed that, using an overall holistic view as his guide, he could generate simple hypotheses that were effective in summing up the parts of a lower level of organization thereby illuminating the properties of the higher level.

That there is a trend toward reductionism in modern population biology is beyond doubt, and one example clearly illustrates this fact. One of the important contributions of W. D. Hamilton (1964) was the notion of inclusive fitness, the idea that fitness should include the survival and reproduction of kin. This means, as was explicit in the early ideas of the population geneticists, that the genes are the object of selection. R. Dawkins (1976) has stated the matter most elegantly in nonmathematical terms in his book *The Selfish Gene*, whose title itself tells the main part of the story. He talks of genes as being "replicators" and the organisms the genes produce through development as being the "survival machines" that are devices for keeping the replicators intact and functioning. We shall discuss this whole matter in detail further on. Here the point I wish to make is that this is evolutionary reductionism in its extreme form, and, as we shall see, these basic ideas have already contributed to significant advances; they are the gateway to a new understanding of evolution and the social organization of animals.

If one looks at the criticisms of sociobiology by anthropologists and social scientists, they are almost entirely related to the idea that a reductionist approach will not be useful in the social sciences. The study of human societies occupies a separate hierarchical level and must be considered in its own terms and not in terms of the biological level lying below. In their view, human societies are too complex, too special, too different from anything found in the animal world to be interpreted in any meaningful way by biological analysis. Their position is largely or entirely holistic. For instance, their notion of culture is that it is an emergent property unique to man. According to M. Sahlins (1976), culture was developed in the hominid line about three million years ago. It is a new condition that came into being as a result of the complexity of the mind of early man. To that extent the cultural anthropologist would consider it biological, but once it came into being, it took on a life of its own, and its new properties cannot be understood in terms of the level below. It is, so to speak, self-propelled and, like a soul, has become detached from its body. More of this later; here I want only to stress that this is indeed a holistic bias. (The very same argument can be made by those, such as J. Jaynes [1977], who consider consciousness also to have arisen suddenly in the early history of man, although consciousness, according to Jaynes, arrived full-blown long after culture.)[1]

It has certainly been true for biology, as I previously illustrated, that when a field is able to make advances by a reductionist approach, the progress is most exciting and rapid. Furthermore it is obvious that the more complex the field, the slower it achieves a stage where it can make fast advances by reductionist methods. This statement is one of simple fact and applies not only to biology, but also to physics and chemistry. And from this I would suggest it is not inconceivable that the same process might occur in the social sciences at some time in the future. If it does, clearly the lower level will be biology. The sociobiologists have already claimed they have

[1] The contrary view is admirably set forth in a book by D. R. Griffin (1976), who with Darwin (1874) provides evidence for the idea that there might be a continuum between what we call consciousness and various manifestations of behavior in animals.

found a bridge connecting the two levels, but this has been stoutly resisted by many anthropologists and social scientists. Again, this is obviously a matter to which we shall return.

One final word on holism. I do not mean to imply holism is bad and reductionism is good, for they are both important. Often in a particular field at a particular stage in its development it is impossible to do anything other than examine the problems holistically. Furthermore a holistic approach has, in many cases, produced significant progress. It is probably true that it is a necessary stage without which the reductionist progress could not be made. Initially, it is the only way of describing the problems and grouping the facts. Were this not done the chaos would be complete. However, despite the strengths of a holistic approach, one should not fear reductionism as an evil. When it comes to a field, it should be greeted with caution, but also with pleasure. The caution is needed because there is a degree of oversimplification where the exceptions may accumulate to such an extent that clearly they no longer prove the rule, but prove the need for a more refined theoretical insight. The more traditional holism keeps the perspective in the field, even when reductionism is rushing forward at a dizzy pace.

It has always seemed strange to me that holism and reductionism should elicit such strong passions among scholars. They are, after all, only the philosophical methods characterizing different kinds of scientific progress. The reductionists tend to be contemptuous of all holists, for they feel they alone have the key to the universe. Holists know they have a broad perspective, a large insight, whereby they can see all the riches missed by the single-minded reductionist. In principle it would appear so easy to be both at once, but human nature is such that it enjoys taking positions on philosophical or political dichotomies, ignoring totally the possibility that some of these dichotomies are not genuine antitheses of the either-or category, but are complementary. In fact, I would go so far as to say that it is the holist who sees and understands the dimensions of the problem and it is the reductionist who in the long run will produce the most satisfying type of explanation. The one cannot do without the other.

A Definition of Culture

There are probably few words that have as many definitions as culture. I can remember when I was a student of Professor William Weston at Harvard, there was a large room across the hall from his office where we made nutrient media to grow fungi and slime molds. It was there that I learned the now lost art of how to make potato agar from real potatoes. On the outside of the glass door to this communal room was one word inscribed in large gold letters: CULTURE.

At the other extreme there are those who use the word in a sense I associate with Matthew Arnold and the *Oxford English Dictionary*: culture is a refinement of tastes and artistic judgments; it is the ultimate in the purification and rarification of the intellect.

Fortunately, definitions in science are arbitrary, and I shall define the word in a sense somewhere in the middle of the great chasm between the two uses of the word mentioned above. By culture I mean the transfer of information by behavioral means, most particularly by the process of teaching and learning. It is used in a sense that contrasts with the transmission of genetic information passed by the direct inheritance of genes from one generation to the next. The information passed in a cultural fashion accumulates in the form of knowledge and tradition, but the stress of this definition is on the mode of transmission of the information, rather than its result.

In this simple definition I have taken great care not to limit it to man, for, as so defined, there are many well-known examples among other animals, especially among those that cooperate extensively such as primates. It would be easy to alter the definition and say arbitrarily that it applied only to man, and since any definition is fair game, there is nothing improper in such a procedure. But I want to emphasize that this is not the course I have taken.

There is a tendency to oppose the words biological and cultural, but Marion Levy has pointed out to me why this is unfortunate. Culture, as I have defined it, is a property achieved by living organisms. Therefore in this sense it is as biological as any other function of an organism, for instance, respiration or locomotion. Since I am stressing the way information is transmitted, we could call one *cultural evolution* and the other *genetical evolution* with the understanding that they are both biological in the sense they both involve living organisms.

Anthropomorphisms

The existence of anthropomorphisms is a problem to which there is no solution. Those interested in the similarities between man and animal have no fear of anthropomorphisms, while those who see man as special in some major way feel that our whole man-oriented language is dangerous and misleading when applied to animals. Here is a clear instance where human culture interferes with our science.

Let us look at the prejudices of both sides of the argument. An anthropologist might find the use of words such as *slaves* or *castes* for ant colonies most undesirable. There are a number of reasons he finds this usage unfortunate. For instance, it implies that the most repugnant human morals are ascribed to the members of some species of ant who are clearly too stupid to be immoral. Much worse, it could imply that if ants have slavery, it is a natural thing to do and therefore quite justified in a human society. These arguments are not quite rational and can only be advanced under extreme fervor of one sort or another. A more reasoned objection would be that the motivations of ants and men might differ radically, but by using the same words this distinction is lost.

A biologist, on the other hand, feels that the points made above are too obvious to interfere with the dual use of the words. He does not see any problem: in both ant and human slavery individuals forcibly capture members of their own species or related species and cause their captives to do work for the benefit of the captors. It is unnecessary to drag in all the possible political, psychological, or strictly human nuances; a very simple definition of the word is sufficient. There is no need to be tyrannized by words. If a biologist may not use the common words, he will be forced to invent a whole new set of jargon terms for nonhuman societies, an unfortunate direction since there are too many jargon words in any science as it is. I hope it will be sufficient if I make it clear in the beginning that words either invented or frequently used for human societies will also be used for animal societies with the understanding that I am not implying anything human in their meaning; they are to be considered simple descriptions of conditions.

There nevertheless is a difficulty. It can be argued that no matter how excellent and pure our stated intentions might be, the words will unconsciously tend to make us interpret animal behavior in human terms. But surely this danger exists no matter what terms we use. It comes down to the very core of the problem of objectivity: we see the world only through our own eyes, our own minds. One might suppose it is easier to separate Newtonian mechanics from our psyche than courtship and altruism in the behavior of birds, but in fact they are both seen through our minds. If anything, in the behavior of birds it is possible to see the pitfalls simply because they are more obvious. The difficulty of attributing human motives (correctly or incorrectly) exists and will continue to do so, no matter how cumbersome a vocabulary one invents. When the reader finds words of common usage in the pages that follow, he is urged to interpret them in a straightforward way. Even if he subconsciously fails in this task, no great harm is done for the question of what motivates other animals, as compared to ourselves, is not the central subject of this book.

CHAPTER 2

Cultural and Genetical Evolution

The ultimate purpose of this book is to explain why we have culture at all. Why is not all the transmission of information from one organism to another fulfilled by genetic means as it is, for instance, in plants. This is a question in evolutionary biology that will be approached two ways.

One approach is to examine those features of culture that are selectively advantageous. Why was there a selection for organisms capable of culture? What is there about behavioral transmission that might make it a useful adaptation in natural selection? The question of adaptive advantage will be touched upon briefly at various points in this chapter and the succeeding ones, but it will be the main subject of the last chapter for it is the climax of my argument.

The other approach to the question of why we have culture at all is to examine the early evolutionary origins of culture, for surely it is not a phenomenon that arose full-blown. I shall show that there has been a whole series of precursor steps, each of which is itself adaptive. It could be argued that culture is the accidental end result of such a series of steps, but as we shall see, many of the steps share certain adaptive properties with culture, even though the steps may seem primitive and far removed from the ability to have culture itself. A description of these steps leading to culture and their possible adaptive significance will make up the bulk of this book. It is an evolutionary progression of exceptional interest and one not often viewed in quite this light.

Before we can begin such an odyssey, there are some general points that must be examined. First we must understand as clearly as possible the distinction between cultural and genetical evolution. That is the subject of this chapter. Next we must examine the nature of the biological structures responsible for these two modes of transmission: we shall look at the brain and the genetic apparatus and see how they both evolved. This will be the subject of Chapter

Three. Chapters Four, Five, [illegible] Seven follow the evolutionary steps leading to culture, and [illegible] Eight, as I said above, examines the adaptive significance [illegible] re itself.

Genetical Evolution

In order to understand the simil[illegible] nd differences between genetical and cultural evolution, it [illegible] sary to have a clear grasp of each. To understand genetical [illegible] n further requires an understanding of natural selection.

The modern biologist's conception of natural selection is deceptively simple. It is that individuals of a species in a population vary. Furthermore, these variations are genetic and are inherited in a Mendelian fashion. Because certain genes and gene combinations will be more successful than others in a particular environment, they will increase in relative number in the population. This is often called reproductive success. Because the genetic constitution of the population will change, the environment itself changes, and this in turn alters the nature of what is selected.[1]

One hidden but essential part of the mechanism of natural selection is reproduction. The selection of genes could not operate unless there were a regular transfer of genes (usually through the sperm and egg) from one generation to another. This succession of generations, called reproduction, is an absolute condition for natural selection.

Because it is crucial to our discussion, it is important to examine biological reproduction in more detail. All biologists understand it, but it is uncommon to think seriously about its role in evolution. I have discussed this at length elsewhere (Bonner 1965, 1974); here I wish only to emphasize the main points.

The organism is a life cycle. It begins with a single cell (a fertilized egg or an asexual spore) and, if it is a multicellular organism, it increases in size by growth with cell duplication and divides the labor among the newly formed cells by differentiation. This is a period of development involving a rapid size increase. Once maturity is reached the individual produces new eggs and sperms that

[1] Genetic changes can affect the environment in many ways. In the simplest terms, different organisms will produce different neighbors. The effect would be particularly obvious and dramatic if the genetic changes involved the social interaction of individuals. The environment of course changes for other reasons also.

start the next generation. Therefore biological reproduction is the reproduction of the whole life cycle, not just the duplication of adults. There are a few rare forms, such as certain species of flatworm, where the adult divides in two and each half regenerates the missing parts, but even in these forms there are also sexual stages where again the cycle comes to a single cell stage, the fertilized egg.

We may ask the question why this is so. Why do even enormous animals and plants like elephants and giant sequoias go through the huge bother of rebuilding an entire organism from a single cell for each generation? The answer to this question has been most clearly expressed in R. Dawkins' (1976) splendid book. He presents W. D. Hamilton's ideas on inclusive fitness in an ingenious fashion. It is the gene, the components of the DNA wrapped in the nucleus of the fertilized egg, that is the important object as far as natural selection is concerned. The other materials in the egg, all the other chemicals and structures, both nuclear and cytoplasmic, are there to nurture the genes, to give them an environment in which they can flourish and in which they can be persuaded to initiate the synthesis of certain proteins that become part of the construction of the body.

If this is so, why then are not all animals and plants simply single cells? It is true that many are, and very stable and successful ones at that. In fact, there are far more microbes on the surface of the earth than multicellular organisms. However, under some environmental circumstances the genes have a good chance of survival if they are carried by large organisms. These no longer compete with the smaller organisms and, by avoiding competition, they can exploit new environments. In effect, they become an efficient means or device for perpetuating the genes. To use Dawkins' terminology, the genes are the "replicators," and the phenotype, the developing organism, is a "survival machine."[2] If genes are to survive, then they

[2] Let me introduce a useful, related set of jargon words here. A *phenotype* is the sum of all visible structures that comprise an organism; it is the survival machine. By contrast, the collection of genes that specify the characters of the phenotype is the *genotype*. This assembly of genes of any one individual organism is also called a *genome*. As I shall show later, genome is a helpful word to use in contrast to *brain*, for the genome contains all the genetic information of an individual, while the brain contains all the behavioral information. The genome is key in genetical evolution, while the brain is key in cultural evolution. The brain is part of the phenotype, and its structure is specified by part of the genome during embryonic development.

must be carried in some safe way. That way can vary enormously from bacteria to higher plants or higher animals. Such variety exists because of the fierce competition in any environment, and, to be successful, the survival machines have to adopt an extraordinary variety of forms. The phenotype is the advance guard that protects the genes. In some cases it proves to be efficient to build enormous survival machines at a huge cost, while under other ecological circumstances the most effective machines may be small. Large machines take a long time to build, while small ones have very short generation times, that is, the time from one generation to the next (Figure 1). Not unexpectedly the mechanisms and rates of gene change also vary with size.

If the phenotype is a survival machine and the genome a replicator, then why is it necessary that there should be a single-celled stage in large organisms? The answer to this question is purely genetic. Consider the case of diploid sexual organisms where it is advantageous to recombine or shuffle the genes at meiosis (that is, in the formation of the sex cells) and to fuse one gamete of each sex to bring together the recombined gene complements of both parents.[3] Since all large organisms are diploid, every cell in the body presumably contains the same diploid complement of gene-containing chromosomes as the fertilized egg: one set from the mother and one set from the father. The selfish genes must ensure that these basic processes of meiosis and fertilization take place, and, therefore, there will always be a stage in the life cycle that is reduced to a single cell. These stages of gene shuffling can only take place once each generation, and they are presumably important to meet the problems of changing environments and new competition. For this reason it is not surprising that population geneticists often think of evolutionary change not in terms of years, but in terms of the number of generations. It means of course that large organisms change more slowly (in absolute time) than small ones, but this disadvantage is offset by the great advantage provided by their large survival machines.

To sum up, the whole life cycle is reproduced. This reproduction is a basic requirement for natural selection. While selection acts upon the phenotype, the net result is a selection for the genes. In

[3] Meiosis is the process in which the cells with two sets of chromosomes (diploid) are reduced to one (haploid). This occurs during the formation of egg and sperm, and when these haploid sex cells fuse in fertilization, they form a diploid progeny.

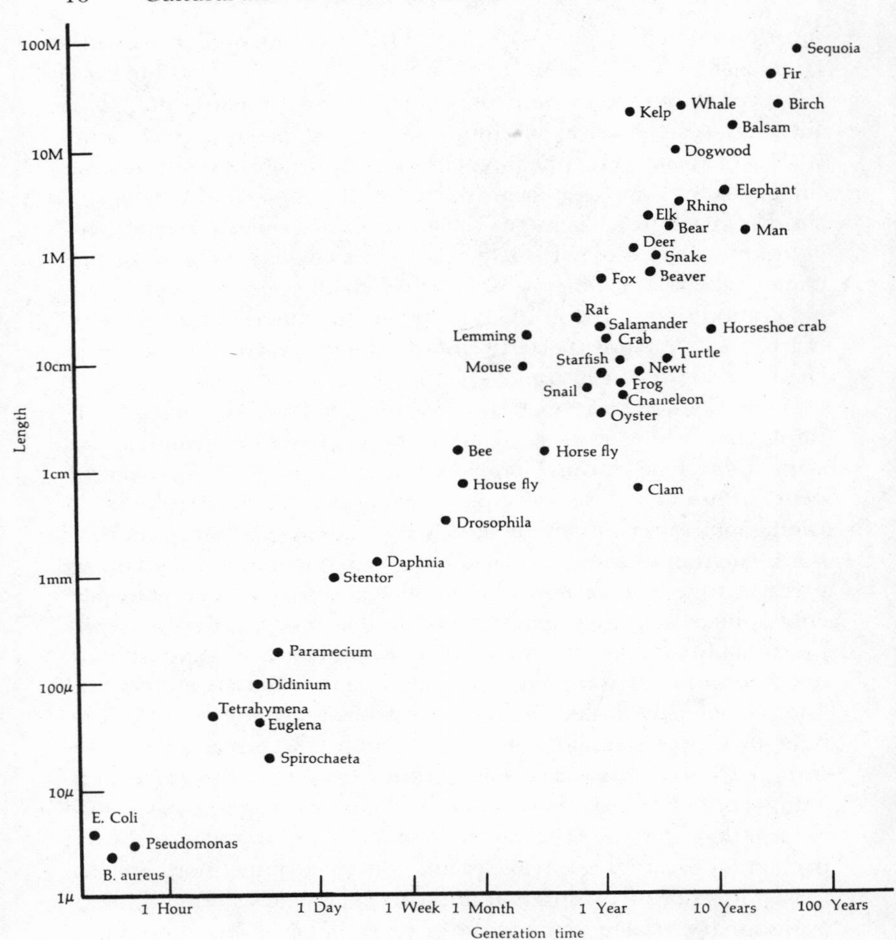

Figure 1. The length of an organism at the time of reproduction plotted logarithmically against its generation time. (From J. T. Bonner 1965.)

turn, these can, and do, construct a great variety of survival machines, large and small. The function of these survival machines, whatever their size, is to protect the genes better in their immediate environment. Since the only time gene changes can occur is during the single cell stage of the life cycle, the rate of evolutionary change necessarily depends upon the number of cycles or generations.

Cultural Evolution Contrasted with Genetical Evolution

As indicated in the previous chapter, by culture we mean behavior transmitted from one individual to another by teaching and learning. If the content of these transmissions changes by innovation or accident, there will be a cultural change or a cultural evolution. To understand the distinction between cultural and genetical evolution, it is useful to use Dawkins' (1976) term *meme*, which is the analogue of *gene*. In behavioral transmission, the transmitted information is a meme, be it an idea, a belief, a custom, or a lesson. Dawkins has not attempted a rigorous definition of a meme, nor shall I here. To keep matters straightforward, I shall use the word in the sense of any bit or any collection of bits of information passed by behavioral means from one individual to another. Therefore in cultural evolution information is transmitted by memes, while in genetical evolution it is transmitted by genes.

The analogy between the two kinds of evolution is compelling and my colleague Henry Horn has shown me that one can formally devise a "fundamental theorem of cultural selection" that precisely fits the mathematical statement of R. A. Fisher's "fundamental theorem of natural selection." In one respect cultural and genetical evolution are identical rather than analogous: they both can be guided by selection. Because memes and genes are affected by the same selective forces, it is often more difficult to see the sharp distinction in any one change between the relatively ephemeral cultural component and the relatively slow responding genetic component.

Perhaps this is the appropriate moment to say that I basically disagree with numerous individuals who find the analogy useful and lump cultural and genetic selection together to arrive at some general principles of change in man. (There is a growing literature in this field. For a review of the views of D. T. Campbell, F. T. Cloak, Jr., W. H. Durham, and others, see P. J. Richerson and R. Boyd 1978.) The main point I make in this chapter is dramatically opposed to such an idea. It is only by making a clear distinction between genetical and cultural change that we shall ever be able to understand the causes and the mechanisms of change in any organism capable of both cultural and genetical evolution.

Therefore I shall stress the differences rather than the similarities forming the basis of the analogy. There are three main differences and, as we shall see, they are closely interrelated.

(1) The most important difference is the one I have already stressed: it is in the mode of transmission of information. In genetical evolution transmission can only occur by the passing of information containing molecules from one individual to another by means of fertilization. The key molecules are primarily DNA, although RNA's, proteins, and other substances may also contribute minor amounts of additional information. The physical restrictions imposed by this kind of transfer of chemical substances are considerable; it means, among other things, that information can only be passed from one individual to another, and any one individual can receive this kind of information only once during its life cycle. For any organism, the number of its offspring equals the number of times it has been able to transmit genetic information, for the transmission can only occur from parent to offspring.

On the other hand, in cultural evolution the transmission can occur by teaching and learning. Any one individual can teach many individuals or can learn from many individuals. This is because the means of communication or transmission are varied and can act over a distance and even over a considerable span of time. The communication can be by visual, auditory, and other kinds of signals, and responding individuals will have special kinds of receptors to receive them. The signals can be stored in various ways, such as in memory, in writing, or in a tape recorder, and in this way be potentially able to elicit change over a long period of time.

(2) One striking and important consequence of this difference in the mode of transmission is that changes in genetical evolution are necessarily very slow, and require many generations, while changes in cultural evolution may be very rapid. If, for instance, a favorable genetic mutation appears in one individual, even with reasonable positive selection pressures, it will take many life cycles and perhaps hundreds of years for the new gene to be present in an appreciable number of individuals in the population. It is only in some lethal circumstances that the genetic structure of a population can change rapidly. For instance, if there are a few individuals in an insect or bacterium population who happen to have a gene that confers resistance to an otherwise lethal poison, then those individuals will be the only survivors and constitute the next generation. Although it is relatively rapid, it is a rather radical way to achieve genetical change.

On the other hand a cultural change can be exceedingly rapid. A new fad or dress style may take over a whole nation in a matter of days or weeks. One could think of many other examples. But there will also be some meme-mediated changes that can be and are constant for long periods of time. Dress styles may change rapidly in Western civilization, but there are many primitive tribes in which there is evidence that the apparel has been the same for centuries. This does not mean, however, that a little bartering and a few glimpses of the outside world will not quickly replace the simple materials of the past with cotton shirts and dresses.

Because of the different modes of transmission, genetic change is always relatively slow, while cultural change is fast or slow. The former is limited by the fact that the information has to be passed from one individual to another, while the latter can, in some circumstances, spread an idea with the same rapidity as the invasion of an epidemic infection, an analogy to be pursued further in a moment.

(3) Memes are utterly dependent upon genes, but genes can exist and change quite independently of memes. Memes are produced by the behavior patterns of living organisms, which reproduce by successive life cycles, each stage of which is controlled by genes. Memes only exist in this context; they are a by-product of the genes. Genes, on the other hand, need not necessarily produce an organism capable of behavior and of manufacturing memes; they could equally well produce a stationary plant. More important, there are many features of a meme-producing animal, such as its skeleton or its circulatory system, that function quite independently of the brain that produces the memes. But all the genes that build the life cycle of an animal provide it with the machinery for its physiological functioning and produce an independent organism that is capable of maintaining a complex brain that in turn is capable of manufacturing and responding to memes.

The meme-making brain is, in the sense discussed above, a parasite of the gene-produced body; it could not exist without the body. At first glance it would seem that all the advantage was on the side of the brain as one would expect from a true parasite, but later I shall make it clear that there are advantages both ways, and the genome may gain by producing a brain. There is a symbiosis between the meme-producing and the gene-producing structures. In fact, this is clearly why we have memes at all: the ability to produce them arose by natural selection because of their advantages to the genome. We shall return to this matter in the next chapter.

If there is such a mutualism, such a reciprocal gain between brain and genome, then it would be helpful to examine the question of how they can affect one another to achieve change.

Can Culture Affect the Gene Pool?

In principle the answer to the question of whether culture affects the gene pool is yes. By culture one can change the environment, and it is the environment that controls the direction of the selection of the genes. As I mentioned earlier, selection can act on both the genes and the memes. If it acts on the memes, there can be two conceivable consequences. One is that the meme itself will persist as long as the selection for it is positive. This selection, by definition, will not affect the genome of the population in any way; it is purely a selection of some cultural phenomenon. The other is that the selection of a meme could ultimately affect the direction of gene changes by favoring reproductive success in certain phenotypes. This is the kind of selection with which we are concerned here.

It is easy to postulate how culture could affect the gene pool; the more difficult problem is to find clear-cut examples. It requires a situation where undisputed cultural change remained stable long enough for natural selection to occur. One might presume this might happen when modern Western civilization has produced a different set of environmental restraints than primitive societies. Various authors (for example, Post 1971) have argued that color blindness, myopia, and deficiencies in hearing acuity show a greater frequency in people who have been civilized and urbanized for longer periods of time than those that have remained in a hunting and gathering state. When various primitive Asian and African tribes are compared to European groups there is, for instance, a clear increase in the incidence of color blindness among the Europeans; but as Bodmer and Cavalli-Sforza (1976) point out, there are difficulties. The rates of change are too rapid in some cases to be accounted for solely on the basis of relaxation of selection. Furthermore, many of these easy to measure characters are sex linked. For color blindness this would mean that in a primitive tribe normal vision would be more important for gathering by women than for hunting by men. This unlikely prediction is because color blindness being sex linked will be rare in women and common in men (as is the case in haemophilia, another sex linked trait).

Perhaps the greatest difficulty with this class of examples, including all those diseases that can now be cured by modern medicine and therefore will not be selected against, is that it is not a positive selection, but a relaxation of selection. Therefore, our best examples of culture affecting the gene pool not only present inherent difficulties of interpretation but they illustrate solely those cases where the change is a cessation of selection. We can only conclude that culture affecting the genetic constitution of a population is theoretically possible without being able to give one good, unequivocal example.

However, one can say something helpful and penetrating on the theoretical side of the relation between cultural and genetical evolution as Feldman and Cavalli-Sforza (1976) and others have shown. They consider the simplified case of a single pair of genetic alleles found in the population in the standard forms of *AA, Aa* and *aa*.[4] But besides these genotypes, the individuals could possess a particular skill acquired through learning. This means there are six possible kinds of individuals: the skilled ones, $\overline{AA}$, $\overline{Aa}$, and $\overline{aa}$; and the unskilled ones, *AA, Aa, aa*. With these players Feldman and Cavalli-Sforza now indulge in a series of games making some special, simple assumptions for each. Among other things they point out that the process of teaching and learning can be treated as though it were an infection, and if the three possible genotypes vary either in their ability to become infected (ability to learn) or their ability to infect (ability to teach), one can test for the changes in both the phenotypes and the genotypes in a population, making certain assumptions as to the selective advantages of the gene combinations and the phenotypes. Even in this oversimplified beginning, the behavior of the culture-genome complex can be examined under a variety of conditions and circumstances. The matter is summed up by Robert May (1977) who says, "formidable mathematical difficulties stand in the way of a more full understanding of the interplay between cultural and biological evolutionary processes. . . . Nevertheless, I believe that the incorporation of cultural inheritance into the quantitative theory of population genetics, as begun by Feldman, Cavalli-Sforza and others, is likely to open exciting new areas" (1977:13).[5]

[4] An allele is an alternate form of a gene. A pair of alleles will occupy the same position on homologous chromosomes of a diploid organism.

[5] Recently Fagan (1978) has proposed a different model suggesting that there could be a positive selection for inventive play in a social group and that the cultural changes generated by the inventions could in turn affect genetical evolution (and its rate of change). As Fagan points out, his model is quite different mathematically from

Such an approach remains pure in the sense that it clearly distinguishes between phenotypic and genotypic changes. Since there is considerable freedom in choosing the rules of the game, one can invent situations where there will be no interaction between the selection of the two types, or one can assume that the selection of one affects the other, or vice versa. Here we are considering the case where the gene pool is affected by cultural changes. In the terms of Feldman, Cavalli-Sforza (or Fagan), it would be a case where changes in the phenotype brought about by cultural skills would affect the selection of particular genotypes.

Let us briefly examine the relation between culture and the phenotype. The phenotype is the survival machine. Its principal characteristics are determined by the genes: during development the genes act and produce the physical structures of an individual organism that is the phenotype. Although the genes may be the underlying determinants of the phenotype, they are by no means the sole ones. The environment plays an essential role as well. The genes act in particular environments, both internal and external, and these in turn modulate the gene actions. In late development, especially in man, culture can be an important component of the environment. By teaching and learning of such things as traditions and moral codes of behavior, the phenotypes may differ greatly in different cultures. This leads to the important point that behavior is part of the phenotype, just as much as the physical structure of the body. Any behavioral pattern is obviously especially susceptible to the kind of environmental influence we call culture. Parenthetically, culture can also alter the body, the phenotype, physically. Examples are legion: circumcision, ritual scars or tattoos, skulls flattened by binding during infancy, and innumerable others. They can even affect reproductive success, the most obvious case being castration to provide a class of eunuchs, so prevalent in the Near East for many years, or to provide adult male sopranos in recent Western civilization. (If musical talent has any genetic component, that certainly was not the way to cultivate those musical genes.)

Feldman's and Cavalli-Sforza's definitions of skilled and unskilled and their comparison of imparting skills with infection assume that skill is entirely a cultural change of the phenotype. Only when they

---

that of Feldman and Cavalli-Sforza, but the conclusions derived from both are in harmony and often identical. (See also an interesting recent paper by Cavalli-Sforza and Feldman 1978.)

postulate genetic differences in ability to learn or teach do genes become involved. The problem is that, if such genes exist (and it is reasonable that they might), they are difficult to measure. So again we find ourselves barred from being able to give concise, hard examples, although there remains no quarrel with the assertion that cultural differences in different populations of a species could affect the gene pool.

Can Genes Affect Culture?

In examining the question of whether genes affect culture it is clear that in one very obvious sense genes do affect culture. Culture is the product of our brain, and presumably all the major physical features of our brain are genetically determined. This subject will be examined in detail in the next chapter when we consider the evolution of the brain.

The more pertinent argument that genes can affect culture comes from the basic tenets of modern sociobiology. There are the views first expressed by W. D. Hamilton and later elaborated by E. O. Wilson, R. L. Trivers, J. Maynard Smith, R. Dawkins, and others (see Dawkins 1976). The main thesis is that while the immediate object of natural selection is the phenotype, the ultimate object is the gene, and in this sense genes are "selfish"; each is out for itself. But if strategies of self-preservation are examined, it is found that one of the most effective ways genes can perpetuate themselves is by having the closely related survival machines that house them be helpful and altruistic to one another. The reason for this is obvious: the genes are not selfish as individual molecules, but as a class of identical copies capable of replication in successive generations. Those genes that do make more than their share of copies are necessarily more numerous in the next generation; this is the sense in which they are selfish. Therefore, if their survival machines are altruistic, even to the point of self-sacrifice, the net effect will be an increase in the number of copies of certain genes in subsequent generations.

W. D. Hamilton made the major contribution to this discussion with his solution of the problem first posed by Darwin: why should social ants, wasps, and bees have neuter workers. These workers normally are barren, and this would seem a dead end for the genes. Darwin (1859) made the correct analogy with the somatic cells of any multicellular organism. He points out that the reproductive cells, the germ-line cells, are the only ones in a multicellular organism that are perpetuated. So all the genes in our bodies thought to be exactly replicated in each of our cells die with us. Only those carried by the germ cells are perpetuated in our children, just as only the germ cells of the male and the queen are perpetuated in the ant colony. This shows that genes are selfish not simply in the sense of making more total copies, but that the copies are in offspring and subsequent generations. Also Darwin's explanation seems to be an affirmation of the principle already discussed that the survival machine may be elaborated to house the selfish perpetuating genes, even to the point of having a whole insect colony be part of the machine.

Hamilton (1964) showed not only that Darwin was right, but why. The genome composition of social wasps, bees, and ants (Hymenoptera) are such that sisters (all workers are genetically female) are more closely related to one another than they are to their mother, the queen.[6] Sharing genes is postulated to be the basis of altruism, the reason why individual worker ants will work so hard to feed, care for, house, and attend their siblings with such diligence. From the point of view of perpetuating certain genes, this is the most effective strategy. Hamilton showed that because the males of Hymenoptera happen to be haploid and females diploid, the relatedness of full sisters increases over most animals and plants in which both parents are diploids. But even in these normal diploid organisms siblings are more closely related (share more genes in common) than cousins or more distant relatives. From this it has been argued that while some social insects form an extreme case (which permitted Hamilton to show the principle), all animals tend to exhibit altruism toward their close kin for the same reasons. This is often referred to as kin selection. To return to Darwin's point, in any one multicellular organism the cells of the body are all genetically related to one another. In fact, they do much better than even the social Hymenoptera, for they are genetically identical. Therefore every cell in the whole body of the survival machine indulges in altruism and thereby helps to perpetuate its genes.

In the case of social insects and all other animals, this altruism takes the form of behavioral patterns in which clearly related individuals come together and in some way provide mutual aid. Let me

[6] A good discussion of Hamilton's paper will be found in E. O. Wilson (1971). For a brief statement see J. Krebs and R. M. May (1976). (See also p. 95.)

give one example thought to illustrate this principle of kin selection. The Florida scrub jay is often assisted by year-old juvenile birds in the care and feeding of its young, as well as their protection from predators (Figure 2). So instead of having just one pair supervising the upbringing of the nestlings, there may often be three or four individuals. G. E. Woolfenden (1973, 1975) has done a study of this situation and shown that if other birds help the parents, then the chances of their raising offspring are substantially increased. Unassisted pairs averaged 1.1 fledglings of which 0.5 were alive after 3 months in 47 cases studied, while assisted pairs produced an average of 2.1 fledglings of which 1.3 were alive after 3 months in 59 cases studied. The data clearly indicate that assistance substantially improves, in fact, doubles, the survival rate. In those cases where it has been determined the helpers are close kin, the study perfectly illustrates the principle of kin selection and altruism in animal behavior.[7]

There is a growing literature on this subject, although there was a considerable lag after Hamilton's key paper in 1964. The recent books of E. O. Wilson (1975) and R. Dawkins (1976) are notable, the first for its encyclopedic nature, the second for its ability to express mathematical ideas in plain words, and both for their clarity of thought. The task here is not to review this important subject, but to see to what extent becoming social is related to the development of culture. The thesis of sociobiology is that social behavior is the consequence of cooperation between kin, this being an excellent strategy to preserve selfish genes. A rationale has been provided to explain a social existence; now we shall consider how this existence is related to culture.

There is an obvious answer to this question. Culture involves the transmission and accumulation of information by nongenic means between members of one species. Clearly this can occur most effectively in a social animal for the very characteristic of such animals is a close system of communication between individuals. All those cases where there is evidence of animals passing information in a cultural fashion (however crude it may be compared to human culture) involve social animals. Good examples can be found among social insects and higher vertebrates, but it is essentially true by definition that one cannot have culture without social interaction.

[7] Woolfenden and Fitzpatrick (1978) also have evidence that the helpers improve their lot in the process, not only by gaining experience, but also by ultimately gaining territories.

Figure 2. Florida scrub jay parents feeding nestlings with the help of a yearling bird (left).

This then raises the thorny question. If social behavior has a genetic basis, can culture? Everyone has agreed that the capacity for culture undoubtedly has a genetic basis; but this could be no more than saying that the genes determine the structure of the brain. Could the cultural acts of man be genetically determined by selfish genes in a way similar to those presumably responsible for the altruistic behavior in the Florida scrub jay? Let us first examine some examples before we see whether or not it is possible to arrive at a conclusion.

The incest tabu is often cited as a case of cultural behavior that is also genetically sound. It prevents close inbreeding and encourages outbreeding. Is there a genetically ingrained psychological aversion to mating between brothers and sisters or close relatives, or is this another case of culture providing a custom that happens to be genetically wise and therefore retained?[8]

Aggression in animals can be shown to have advantages in the competition among genes and therefore is thought to be largely a genetically determined phenomenon in birds and nonhuman mammals. This is a subject that has been ingeniously examined by J. Maynard Smith (1976), and the theoretical arguments for a genetic basis are convincing.[9] Is there any possibility that human aggressive behavior, of which examples are not difficult to find, could similarly be entirely the result of rivalry between genes? Other clever notions concerning human behavior have been advanced, such as R. L. Trivers' (1971) theory of reciprocal altruism, in which he suggests that altruistic individuals could count on receiving equal aid in return.

If one restricted the discussion to theoretical grounds and asked could the incest tabu, certain forms of aggression, and mutual aid be largely genetically determined in a human population, then the

[8] It is generally assumed that the mechanism of incest avoidance is due to the more general avoidance of sexual activity between individuals who have been reared together. Shepher (1971) made the interesting observation that unrelated children raised in the same kibbutz do not marry. Presumably if there is a genetic component to the incest tabu, it acts in this manner. It should be noted here that a balance must be struck between excessive inbreeding that is favorable for kin selection and outbreeding (encouraged by the incest tabu) that is a key element in the successful evolutionary role of sexual reproduction. As Wilson (1976) points out the problem of how this balance is maintained is a central one for sociobiology. For an excellent recent discussion see R. M. May (1979).

[9] See also Dawkins (1976: 74ff.) for an excellent review of the ideas of J. Maynard Smith and his collaborators, G. R. Price and G. A. Parker.

answer is probably yes. It is the merit of sociobiology to show that certainly much social behavior in animals can be explained this way. But if one were to ask the question of whether or not these same patterns of behavior in humans could theoretically be totally determined by nongenetic means, the answer would also have to be yes. Clearly man does many more things by independent, imaginative decision than do animals, and there is no reason the above three examples could not be placed in this totally cultural category. If they turn out to be genetically advantageous as well, then so much the better, but this is not a necessary requirement.

But theoretical arguments here are not enough, and, unfortunately, it is not clear that one will ever be able to determine to what extent any human action is genetically or culturally determined. We come back to the same problems involved in measuring the nature or nurture content of intelligence; at the moment there is no precise way one can test the possibilities. One can make hypotheses and test them to see if they are theoretically sound, but that is not the same thing. Some of those hypotheses will be appealing for one reason or another, but they will always have to be approached with caution.

Everyone seems to be aware of this particular problem and the difficulty in resolving it, and I want to avoid all religious and political biases. The sociobiologists themselves take the position that there is a combined genetic and cultural contribution, and that if one is looking for the origin of certain behavior in early man, then surely some of the behavior of lesser vertebrates could be reasonably expected to apply. Many anthropologists and social scientists, of whom M. Sahlins (1976) is an eloquent representative, take the view that culture can have no direct genetic determination. It is a new form of social by-product that arose in our early history, and once launched, it became an entity of itself. It was and is governed by its own inner rules, and the simple-minded view that there could be any genetic component is seriously misguided. To support his point Sahlins presents some most interesting cases among primitive societies where all the kinship rules expected from the pure Hamiltonian inclusive fitness argument are defied.[10]

In my own view any hypothesis that includes the possibility of

[10] Sahlins' (1976) thesis has been strongly contested by other anthropologists who are more in sympathy with the possibility that the study of man could gain from the advances in sociobiology. See, for instance, Etter (1978).

both genetic and cultural components is the most desirable. The arguments for kin selection, and the evidence it does indeed apply to animals other than man is compelling. The strongest support has come from a paper of Trivers and Hare (1976) who show that the sex ratios in different groups of social insects correspond well with what one would have predicted from kin selection theory.[11] While there is still some dispute over this paper (Alexander and Sherman 1977), it is generally accepted as strong confirming evidence for kin selection. The argument has been applied to many cases in higher vertebrates. We have already mentioned the Florida scrub jays; let me briefly mention another. B.C.R. Bertram (1976) showed that African lions have prides of individuals centered around a number of males that are usually brothers. Occasionally, however, there is a fierce fight for ownership of the pride by marauding bachelor males (also closely related), and if these new males are successful in chasing away the defending males, they will also systematically kill a significant number of the cubs (Bertram 1975).[12] The kin selection argument is that they, as gene survival machines, wish to perpetuate their own genes. It is indeed an effective way.

The cultural anthropologist sees such extraordinary diversity in the people of the world that to him it is inconceivable that such a simple mechanism could account for either the differences among the cultures or even the varied aspects of any one culture. The social scientist is the holist who sees the dimensions of the problem; the sociobiologist is the reductionist who works to understand one of the facets of culture. That both views should be tolerated and embraced seems to me to be without question. The only intolerable position is that of either extreme: that all or no cultural phenomena have a direct genetic involvement. The difficulty is the one we began with, namely to know how to separate the two. At the moment, and I suspect for a long time to come, we shall be dependent upon the theorist who can give us not only insights into the possible consequences of different mixes of gene and environment, but who may also come up with clever explanations not intuitively obvious. This, after all, is precisely what Hamilton did, and the illumination he has produced has been nothing short of spectacular.

[11] For a brief, clear summary of this paper see Krebs and May (1976).

[12] There are other examles of infanticide by marauding males invading a social group, as, for instance, in some monkeys, more particularly, langurs (Hrdy 1978).

CHAPTER 3

The Brain and the Genome

We have compared cultural and genetical evolution; now the stage is set to compare the two physical bodies responsible for those evolutions: the brain and the genome.

They are both information processing structures. The brain processes thoughts, movements, immediate reactions to the environment, in sum all the activities we associate with animals. The genome processes genes by replication, and the genes are responsible for making specific proteins that in turn are required to build the structure of the organism through its entire life cycle. The basic similarity between the two is that they both take in, store, and give out information; the difference between the two is not only that the information differs, but that they are on a different time scale. Reactions of the brain and the nervous system are rapid, while those of the genome are, by comparison, exceedingly slow.

The Brain as a Symbiont of the Genome

As has been repeatedly pointed out, the brain itself and all the attendant peripheral neurons are products of the genes. That is, in each life cycle, the structure of the nerve cells, their position, and their arrangement are essentially results of gene actions. Therefore, during the course of genetical evolution, selection pressure for the machinery must have produced the fast information processing of the nervous system. This ultimately led to a progressive centralization of the neurons (still by natural selection of the genes) leading to larger brains and finally to those capable of inventing culture. This evolutionary progression is the backbone of this book. Here I want to point out that one information processing machine (the genome) has spawned another (the brain). Furthermore it has created a machine that can process information in new and different ways, the most striking of which is the difference in the rate of processing. The slow genome has, over millions of years, given birth to the rapid brain.

Since they both process information, the evolved brain can now provide some of the processing that had been the sole province of the genome. For example, the only way some primitive animals can adapt to extreme low temperatures is by the slow appearance of genetic variants that develop more fur or more blubber; but in man the problem is solved in a flash by donning heavy clothing. In this instance the brain has bypassed the genome. It has produced the necessary condition for survival, that is, the insulation, and it has done so without the cumbersome process of gene changes involving many generations. In one view the genome has produced a monster that can now respond to selective forces with lightning speed; yet it is theoretically a benevolent monster in that it responds to the same selective forces as the genome, but does so more rapidly and efficiently. It is in this sense that I would call the genome and the brain symbionts. Each depends upon the other for benefits. The brain is an extraordinary device that makes adapting to environmental changes quick and easy, but it is utterly dependent upon the genome for its existence. The brain therefore needs the genome, and the genome gains by the agility of the brain to adapt quickly. Or to put the matter in Dawkins' terminology, the brain makes for a particularly efficient survival machine, and in this way helps in perpetuating the genes. At this point one might ask why is it that all organisms do not have brains. The answer is basically that there are different ecological niches, and some are successfully exploited by brainless organisms, just as there are separate niches for both large and small organisms. But this is a matter to which we shall return in detail later for it comes to the heart of the problem of the evolution of brains and ultimately culture.

Functional Comparison between Brain and Genome

Having made my basic point, I should like to examine in more detail the brain and genome in terms of how they function. They are both information processing devices, one fast and one slow, and the kind of information produced is of a radically different nature. By information processing I mean that, like a computer, they take in certain information, and alter this in some way so as to emit a specific set of instructions.

In the case of the genome, the information is encoded on the DNA that makes messenger RNA by a chemical template system; this messenger RNA wanders into the cytoplasm of the cell, attaches to the ribosomes, and ultimately specifies the amino acid sequences of particular proteins. The question of how those proteins can become instrumental in shaping the whole life cycle of a complex multicellular organism is the central problem of developmental biology.

One important aspect of the genome is the fact that the genes interact in various ways. There are regulator genes that control the activities of numerous structural genes in simple and complex ways; not only do the genes affect the expression of other genes, but the gene products do as well. The question of why a particular gene is transcribed (that is, makes messenger RNA) at a particular moment is presumed to depend on the immediate environment of the genome, which will vary in time and in different parts of the multicellular organism. Over and above these intricacies, many genes are pleiotropic, that is, they have more than one effect on the structure of the developing organism. A gene affecting eye color can also have some less obvious effect on the construction of the kidney or the nervous system. Genes and their products form a complex, interacting system, but these interactions are not haphazard. Quite the contrary, they are highly structured and ultimately responsible for the orderly process of development.

In the brain, part of the store of information is engraved permanently and arises during development in each cycle. The rest of the information results immediately from signals received from the sensory system: our eyes, ears, and organs sensitive to chemicals and touch. The former is the direct inheritance of a pattern of behavior. It is not entirely clear how the genes act to produce this behavior, but I shall say more on this later. Here we are concerned with the direct reading of environmental information and transforming this knowledge into appropriate actions like the avoidance of something too hot or the seeking of something that smells delicious.

There is a continuous progression of complexity as one goes from a reflex to a conscious decision of whether or not to reach for the jam pot. In the first only one neuron may be involved, and the impulse spreads from one branch of the nerve axon to another. The next primitive step upward involves two neurons: a sensory and a motor neuron. The stimulus (which depends upon what activates the sensory neuron) directly acts on the motor neuron, which might cause the contraction of a muscle. From here upward the plot thickens. In the first place neurons can be excitatory or inhibitory, and therefore stimulating a neuron can have one of these two results. Secondly, and most important, the interneurons begin to play an essential role. They connect the sensory and motor neurons in all sorts of complex ways, so that a sensory input can be magnified or reduced. Furthermore, different sensory inputs can be integrated to give the appropriate response. When we eat, the palatability of food is dependent upon what we see, smell, and taste. The great increase in the number of interneurons is the factor leading to our complex brain. How the brain works is of course one of our most important problems in modern biology. But there is a direct communication between neurons, and this has led to the modulating, coordinating, and transforming of information in quite an extraordinary way.

Not the least of the marvels is memory. This is somehow managed by great masses of neurons that lie in a particular region of the brain. W. Greenough (1975) has found that learning increases the connections between the neurons of the cortex of the brain. Young rats taught specific tasks had far more cortical interconnections than an underprivileged control group. It had been known for some time that learning produced new RNA and protein synthesis; now it is presumed that this is part of the construction of the new connections and not, as some of the earlier workers thought, because the molecules themselves contain the learned information. The idea that using the brain, at least at some early, critical stage in its development, will increase its complexity is quite in keeping with reactions to the use or disuse of other organs. For instance, a muscle will increase (within limits) if it is constantly exercised. Genetical evolution has produced an extraordinary structure in the form of a brain: it not only can instantly manage many adaptations that take the genome thousands of years, but is capable of inventions, memory, and therefore cultural evolution. The brain and the genome are mutually dependent; it is a true symbiosis.

Genetics of Behavior

Above I pointed out that genes affect the brain in two directly related ways: one is by brain structure and the other is by the direct inheritance of patterns of behavior. These structure-based patterns are sometimes called instinctive or innate; they stand in clear contrast to those behavior patterns that are flexible and, as one goes up the scale of organisms of increasing brain capacity, ultimately lead to learning and inventing.

Let me give an example of an inherited behavior pattern. The North American cowbirds, like the European cuckoo, are parasitic birds. They lay their eggs in the nests of other birds and in that way completely forego the pleasures and chores of rearing offspring; the host bird does it for them. A. P. King and M. J. West (1977) have reared individual female cowbirds in isolation and shown that, after they mature, a female will respond to the song of a male by immediately adopting a "copulatory posture" (Figure 3). They will not give this response to the songs of other species (and for some unexplained reason they respond with greater enthusiasm to the song of males that have also been raised in isolation). There does not seem to be any way the female cowbird could have learned how to respond; there is every reason to believe that the response is innate and therefore genetically determined.

This immediately raises the interesting question of how the brain can be born with an automatic ability to respond in a specific way to a specific song. One presumes that something in the auditory part of the brain, when it receives a particular sound signal, auto-

Figure 3. The copulatory posture assumed by a female northern cowbird in response to the song of a male. (From a photograph in A. P. King and M. J. West 1977.)

matically sends a volley of impulses to those parts of the brain that regulate reproductive behavior. The assumption is that it is a special pattern of neuronal connections that can respond in this way, and that these are directly established by the genes during the development of the brain. This whole genetically determined transmission of song involving both the singing by the male and its reception by the female is in sharp contrast to the song transmission of many other birds where learning is also involved.[1]

The selective advantage of such a genetically determined pattern is obvious. Parasitic birds do not have a normal home life; they do not build a nest or have a nest-oriented territory. When a female is sexually receptive, it is essential to find a mate. In these rather casual social relations, an auditory signal is clearly advantageous, but it cannot be a learned one, for the birds would be attempting to mate with the males of their foster species. A young female cowbird or cuckoo may never hear a male sing until it is ready to mate; therefore, it must have some system of immediate recognition. It would be interesting to know what property of the female bird attracts the male, because the process must work both ways.

This example of the cowbird is useful in that it is an extreme case of a behavior pattern directly inherited for good selective reasons. There are far more instances where the behavioral character is formed in part by the genome, and in part by the environment, especially by learning. There is even the interesting possibility that the actions of some animals might be to some degree self-taught. This subject has produced a vast literature, and much of it is concerned with the difficulty in separating the learning and the genetic components in any particular behavioral act. Like so many things, there is most likely a continuum from the extreme genetic determination of song reception found in the cowbirds to the opposite extreme of imagination and creativity in man. In general, ethologists have in recent years been successful in avoiding this ancient nature or nurture problem. This does not lessen the interest in our understanding of any particular example, such as the cowbird. In a recent paper P.P.G. Bateson (1975) reviews the controversy from what we know today and points out that part of the confusion can be dispelled if we separate determinants (genetic or environmental) that have specific effects from those that have general effects. An

[1] See J. L. Brown (1975) for a good summary of the large literature on bird song and the role of learning and inheritance.

example of the former would be the cowbird song or, for learning, the specific trial and error discovery by a bird that a particular insect is distasteful and is henceforth to be avoided. The general effects might be a genetically determined response to light in an insect or an environmentally determined reaction to crowded conditions by migration in locusts. If one now assumes that there is both a continuum between the specific and the general, as well as the behavioral response that is largely genetic versus one that is largely determined by environmental circumstances, then one can get a full measure of the possibilities.

Perhaps most important of all is to remember, as the pioneer ethologists K. Lorenz and N. Tinbergen emphasized so forcefully, that the selective advantages of any particular behavior is paramount. In some cases, and again the female cowbird's response to the male's song is an ideal example, the behavior must be largely genetically determined, for the reasons already given. In others, such as the discrimination between different color and shape patterns of insects by birds, learned behavior will be more effective. It will permit the bird to choose from among the myriad of insects available in different combinations at different seasons and to find those that are palatable. To fix such knowledge in the genome would be an infinitely more complex and far less efficient way of handling the same problem. There will, however, be many cases where as far as natural selection is concerned, it is immaterial whether or not the behavior is primarily genetically or environmentally determined; both work equally well. It is even conceivable that in some cases behavior is determined both ways simultaneously as a kind of double assurance to retain some especially desirable behavior pattern. Another possibility is that it is, in some way, partially genetically determined and partially environmentally determined. This is the case of the song of many nonparasitic birds. They will inherit a basic song, but in the presence of parents and close neighbors, they will learn not only to give a fuller, richer song, but to master their local dialect, which is entirely a learned character of the song.[2]

Bird song is a useful example because it exhibits such diversity in the extent of the genetic and the learning or environmental component. Among insects and other invertebrates one finds the best

[2] The work of P. Marler and others. See J. L. Brown (1975) for one of the numerous reviews.

evidence for fixed, inherited action patterns. Consider, for instance, a solitary wasp. The female deposits her eggs in small cavities, adds some food, and seals off the chamber. Upon emergence the young wasp has never seen one of its own kind, yet it can walk, fly, eat, find a mate, mate, find prey, and perform a host of other complex behavior patterns. This is all done without any learning from other individuals. It is awesome to realize that so many (and some of them complex) behavior patterns can be determined by the genes. Behavior of a surprisingly sophisticated nature can be built into the system by instructions from the DNA.

We return now to the all important developmental question of how this is managed. Presumably specific behavior patterns result directly from particular circuits of neurons. If a series of neurons are connected in a set way, this somehow ensures that a specific action will result. In those invertebrates with large neurons, such as molluscs, it is possible to test the function of some of these neurons by stimulating them with an electrode. In a particular ganglion each cell can be identified and numbered, and for each its position and function known. Some are stimulatory neurons, some inhibitory, and others modulate the action of neighboring neurons so that a particular motor action, such as feeding, can be accomplished by the right sequence of muscular movements. The pattern, the geometry of the distribution of neurons, will be the same for one ganglion in each separate individual of a species. In other words, the anatomical pattern is fixed and inherited, and gives rise to the fixed behavior pattern.

It is assumed that learning works for the same goal, namely to shape a particular neuronal pathway, and once this has been achieved, the lesson has been learned. The evidence for this comes in part from the work of Greenough (1975), mentioned earlier, in which young rats that had undergone a rigorous training program were compared anatomically with some deprived of the benefits of education. The trained animals showed a large increase in the number of connections between the neurons in the cortical regions of the brain. It has not been possible, as yet, to show such changes in invertebrate systems, which also are quite capable of a simple learning. There is a reason why it might even be impossible to demonstrate. All the morphological connections between neurons might be present, but only some connections might be operational. Learning could then involve the reopening of certain connections and the closing of others. The resulting physiological change might have no gross anatomical effect, although perhaps it could be detected by electron microscopic analysis of the synaptic connections.

This discussion so far lacks one particular dimension, namely the magnitude of vertebrate brains, especially that of man. We have considered the fixed anatomy of the ganglion of a mollusc containing one or two dozen cells; the human brain has over 10^{10} neurons. Furthermore, each neuron in the cerebrum has between one thousand to ten thousand connections or synapses. These are staggering figures, large enough to make it seem reasonable that we are endowed with enormous powers of memory, inventiveness, imagination, and various less noble mental abilities. The first really genuine insight into how a complex brain of a mammal works comes from analyzing the supposedly simple question of how we record what we see. This has been illuminated by the exemplary researches of D. H. Hubel and T. N. Wiesel (1963), and no doubt in the future we shall have a clear understanding of other brain functions as well. It is an exceptionally interesting and important subject that has drawn the attention of many scientists.

Evolution of the Brain

Since I am concerned here with the forces of selection that modified the genome to produce the brain, let me make a very rough survey of some of the major steps involved in the evolution of the nervous system. We shall begin very simply by assuming a selection pressure for quick responses to environmental changes and further that neurons, both sensory and motor, have already evolved.

The first major step toward culture is the centralization of the nervous system and the formation of a brain. Why would such a move be selectively advantageous to some invertebrates? The answer may be a simple one of mechanical efficiency. If the nervous system of an animal is to increase, either because the animal is evolving to a larger size and needs more neurons to control its parts or because it is advantageous to sense the environment more critically and respond to it in a more sophisticated fashion, centralization is probably the most effective method to adopt. The basis for this contention is that a brain or a ganglion enables the interconnections between the cells to be shorter. As my colleague W. G. Quinn has pointed out to me, this will not only mean less manufacture of cytoplasm for the connecting fibers, but more important, it will permit the responses to be quicker because they travel over shorter routes. Furthermore, because of the proximity of the cell bodies, it would be increasingly easy to construct complex interconnections between them if they are only separated by short distances. The point is set in perspective by comparing a brain with a nerve net such as that found in jellyfish and hydroids (Figure 4). They can contract at any spot that has been stimulated and produce a wave that spreads indiscriminately in all directions; the control is evenly distributed over the whole body of the organism.[3]

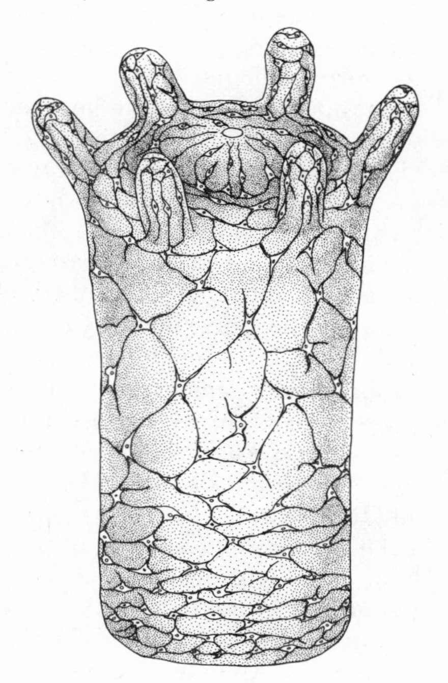

Figure 4. The nerve net of a contracted hydra. (After J. Hadži 1909, *Univ. Wien Arb. Zool. Inst.* 17.)

[3] In a most interesting experiment Campbell et al. (1976) produced strains of hydra that lacked nerve cells completely (and subsequently have been able to perpetuate these nerveless individuals by force feeding them for eighteen generations). The nerveless hydra differed from the normal ones in three significant respects: (1) they showed no spontaneous activity, but would stay essentially motionless most of the time; (2) they could be stimulated to contract, but a much stronger stimulus than normal was needed in order to obtain a response; (3) the wave of contraction was far slower than normal. In other words the ectoderm could take over the second two functions, but with significantly less efficiency than when the nerve net is present. But the ectoderm is quite unable to generate any significant spontaneous contractions.

The role of centralization is beautifully illustrated in the nervous system of segmented animals like annelid worms or arthropods. In the first place there may be a certain amount of rhythmic activity, such as the sequential contraction of the muscles in the segments or the movement of the swimmerette or gills. But rhythmic contractions are also characteristic of jellyfish; the difference is in the way they are controlled. In each side of each segment of a crayfish there is a ganglion, and each of these ganglia is connected to its posterior neighbor. Such a segmental ganglion will not give directions to a swimmerette to move unless it receives the information via the *coordinating fibers* from the ganglion anterior to it that has already contracted. Therefore, the fact that the swimmerettes contract in the proper sequence from anterior to posterior is entirely due to the coordinating fibers connecting the ganglia (Figure 5). This part of the system closely resembles a nerve net; the difference is in the presence of the ganglia that control the relatively elaborate movements of the appendages. But the crayfish also has a participating brain. There are so-called *command fibers* extending from the brain to each of the ganglia, and these fibers control the rate of movement: the time interval between the contraction of one appendage and the next. In sum, the ganglia with their coordinating fibers assure that the swimmerettes move in the proper fashion and in the proper sequence; the brain controls how fast the animal moves. Both processes are essential for the well-being of the organism, and there is a perfect division of labor between the coordinated ganglia and the brain.

Therefore we must think of primitive centralization in both the brain and the ganglia as a means of controlling and coordinating movement, of translating information received by the senses from the environment and transferring it into the proper muscular movements. And, as was pointed out earlier, each neuron and its exact location in all parts of the crayfish is presumably genetically determined, like every other part of the crayfish anatomy.

If one views the invertebrate phyla, one can see an enormous variety in structure and locomotion. This has resulted in equally numerous variations in the construction of the nervous system; many structural designs of the coordinating system are possible.

The next big step, which occurred in many of the more advanced invertebrate groups, was to add learning to the repertoire. This is another way to respond to the environment, and it opens up all sorts of new possibilities. Learning can be used as a way to benefit

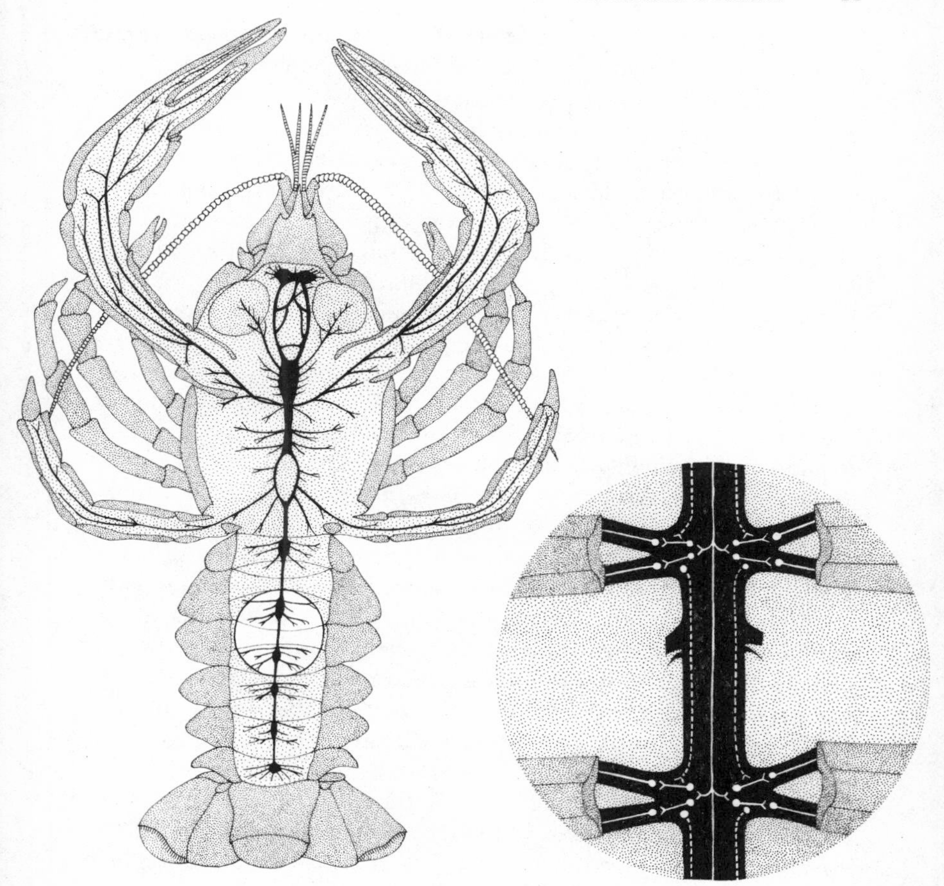

Figure 5. The nervous system of a crayfish. The enlarged portion shows a rough diagram of the command fibers represented by the solid white line down the center of the nerve cord and the coordinating fibers represented by the dashed line on each side of the nerve cord. The orientation of the neurons to the swimmerettes is also shown. The coordinating fibers control the sequence of the muscle movements of the swimmerettes; the command fibers, which are from the brain, control the rate of the sequence.

immediately from trial and error. I have mentioned its advantages in food selection among birds; the same principle would apply to other aspects of the existence of animals. Hymenoptera, for instance, are good at memorizing mazes. Clearly if digger wasps can remember where their nest is located, this vastly increases the chances that they will get back home after a foraging expedition.

Most remarkable of all is that insects, like humans, have both a long term and a short term memory. This has been shown by R. M. Menzel et al. (1975) in honey bees, using various kinds of anesthesia. In another interesting study W. G. Quinn and Y. Dudai (1976) showed that cooling fruit flies immediately after training blocks the subsequent expression of learned behavior, while cooling them thirty minutes later does not. Memory is of course one of the important elements that permits culture, and it is not surprising, as we shall see later, that insects have a rudimentary culture of their own.

The selective advantage of larger brains is evident in the evolution of vertebrates. There is a direct inverse correlation with the time of appearance of a group in earth history and the size of its brain. At one end of the spectrum fish have small brains, while at the other end mammals have the largest. This suggests a trend toward increase in ability to learn, toward an increase in the flexibility of the response. Note, however, that this expansion of the brain probably corresponds largely to the expansion of new niches, and not solely to the elimination of animals with smaller brains. In the early Ordovician era there were only fish. Today fish remain but in addition there are amphibians, reptiles, birds, and mammals. Undoubtedly the presence of these other groups has affected the kinds of fish that exist today (avoiding predatory penguins, alligators, seals, and dolphins has led to special protective adaptations in fish); but there are still fish, and they are abundant and successful as a group despite the relative insignificance of their brains.

In calculating the average size of the brain in any particular group of animals, we find that one of the most important factors affecting brain size is the body size of the individual animal. Big fish have big brains, and little fish small brains. It s possible to plot the logarithm of the brain size and the logarithm of the body size of any group of animals to show this relation clearly (Figure 6). The straight line relation between the two is known as allometric and follows the simple formula:

$$y = bx^{a}$$

or

$$\log y = \log b + a \log x$$

where y is brain weight and x body weight, a is the slope of the line, and b is a constant that gives the value of x when $y = 0$.

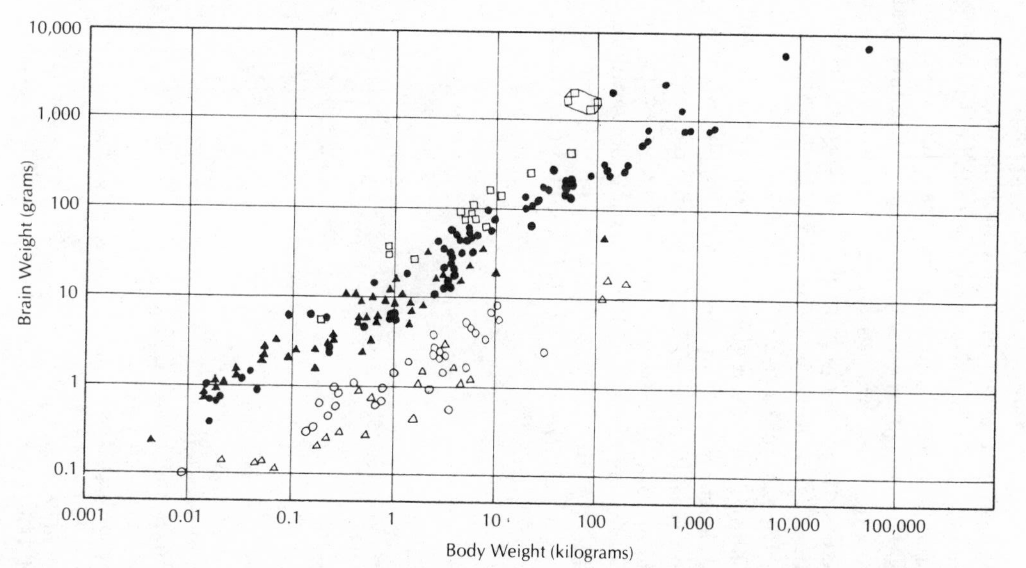

Figure 6. Brain size of 200 species of living vertebrates plotted against body size on a logarithmic scale. Open squares are primates, with man included in the circled box of 4 squares. Mammals are solid dots, birds solid triangles, bony fish open circles, and reptiles open triangles. (Redrawn from H. J. Jerison 1973.)

Does this mean that larger animals are more intelligent than smaller ones, and explain why, in some instances, there might have been a general selection for size increase? This is a very difficult question to answer. First, it is understood that larger size requires more muscles and more nerves to control the movement. So an appreciable component of the brain size increase could be due to just keeping up with the body size. Also note that the slope of the line in Figure 3 is less than one (0.66 or 2/3), so that as the animal gets larger, there is relatively less brain. Despite these two points, the idea that increased body size could improve brain power has been put forward seriously by B. Rensch (1956, 1960). He has shown by comparing similarly constructed animals, such as a bantam chicken, a standard-sized individual, and a giant strain, that not only could the larger forms learn more, but they could retain what they learned for greater periods of time. Rensch became particularly interested in elephants and indeed has shown that, as every child knows, they never forget. Despite his interesting observations it would be very difficult to demonstrate that this is one of the reasons for the selection of overall size increase. It is an intriguing possibility, but no more than that.

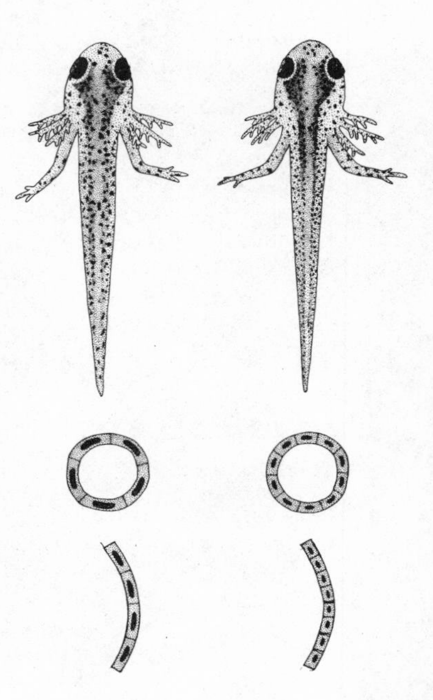

Figure 7. Two larvae of the salamander *Triturus viridescens*. The pentaploid larva with five sets of chromosomes (left) is the same size as the diploid larva with two sets (right). As can be seen from the circular cross section of a kidney tubule in the middle and a covering layer of cells over the lens, the tissues also are the same size despite the striking differences in cell size. (From photographs kindly supplied by Dr. G. Fankhauser.)

As an aside, an interesting experiment was run at Princeton University by a senior student W. Slack with G. Fankhauser (Fankhauser et al. 1955) a few years ago. Fankhauser (1945) had shown in a classic study that in salamanders the triploid individuals (individuals with three sets of chromosomes rather than the normal two) were the same size as the normal diploid ones, while their cells were larger and less numerous (Figure 7). Slack undertook to examine the learning ability of these two kinds of individuals in a Y maze and found that the triploids are significantly less able to learn than the diploids. This means that it is not the total mass of neuronal protoplasm that is critical, but the number of cells. Therefore a large brain can only have advantages provided the large size is associated with more neurons.

A valuable study of the evolution of brain size has been made by H. J. Jerison (1973, 1976). If one plots various groups of vertebrates on a log-log plot as in Figure 6, then one can see that while the slope of each is roughly the same, they tend to lie in two clusters, one below the other. This means that birds and mammals do have relatively more brain than fish and reptiles, and primates manage even better, and this increase is independent of body size.[4] It must be remembered that these measurements always refer to adult brain size. Presently we shall examine how the brain size changes with respect to body size during embryonic development.

[4] Note that since in Figure 6 the different groups of vertebrates lie in two parallel lines, the relative difference between the groups is reflected in the constant b in the allometric equation $y = bx^a$ or (brain size) $= b$ (body size)$^{2/3}$. Jerison (1973) has developed a way of comparing brain size independently of body size by using the ratio of the actual brain size of a particular animal to that expected from an average animal of equal weight from the same taxonomic group (for example, mammals). He calls this the encephalization quotient or EQ, and it can be expressed more formally as follows. (The value of b^* is that of an entire taxonomic group; for mammals it is 0.12.)

$$EQ = \frac{\text{Brain size of a particular mammal}}{\text{Average brain size of a mammal of the same body size}}$$

$$= \frac{b \times (\text{body size of animal})^{2/3}}{b^* \times (\text{body size of animal})^{2/3}}$$

$$= \frac{b}{b^*} = \frac{b}{0.12}$$

For a recent study of the evolution of brain size see L. Radinsky (1978).

Evolution of the Brain in Man

There is much literature on both the embryonic and evolutionary development of the human brain, and much of it is admirably summarized in S. J. Gould's (1977) recent book. Here I do not intend to give an historical account of any of these subjects, but wish merely to isolate what seem to me the key points and to present them as clearly as possible.

First and foremost is the concept of neoteny. By neoteny I mean the retention of certain juvenile characters of the developing organism in the adults of descendants.[5] As Gould shows in detail, the idea that man is a neotenous ape goes way back and has been discussed by a large number of authors. The fact that many of the anatomical features of adult *Homo sapiens* bear a close resemblance to features found in young foetal apes is the basis of Desmond Morris' *The Naked Ape*. Besides hairlessness, there is our flat face, the form of the external ear, small adult teeth and their late eruption, and many other traits including of course a high brain weight relative to body weight.[6] This is nowhere more strikingly shown than in an illustration of A. Naef reproduced on page 354 of Gould's book (Figure 8).

In man there seems to be an extraordinary prolongation of youth, and this permits the brain to continue its expansion. The period of dependency on parents is increased so that the period when learning can occur becomes relatively long (Figure 9). These facts are commonplace knowledge and provide a sensible and insightful way to see the differences between man and ape.

To obtain a better view into this matter it may again be helpful to use the allometric, log-log graphs. It has already been shown (see Figure 6) that if different species from the same major group are plotted on such a graph, a rough straight line results showing that

[5] I differ from de Beer (1940) and Gould (1977) in that I do not cling to the complex terminology that represents two alternatives: (1) the descendant became sexually mature at an earlier stage and therefore the juvenile characters are retained or (2) the development of the body has slowed down with respect to the point of sexual maturity. One reason for not being anxious to give these each their own label is that not only is it hard in any one case to distinguish between the two but that in many cases a little of both could be involved. It seems to me that the more important principle is that of *heterochrony* that says that different structures of the organism, including the gonads, can appear at different times relative to one another during the course of development. For further discussion, see Bonner (1965).

[6] These are the data of L. Bolk. See p. 356 of Gould (1977).

Figure 8. A comparison of the facial features of a baby (left) and an adult (right) chimpanzee. (Drawn from photographs of A. Naef in S. J. Gould 1977.)

brain weight increases less rapidly than body weight; the slope of the line is about 0.66. It is also well-known that if one compares different-sized, closely related species, the slope of the line is very much less (between 0.2 and 0.4). In other words, the advantage gained for brain size increases from increasing body size in any one small group is even less than in relatively unrelated forms. To illustrate this point, I have used a graph from D. Pilbeam and S. J. Gould (1974) in which it can be seen that the brain-body weight relations for the great apes fall on a line with a similar slope of 0.33 although *Australophithecus africanus* already starts with a distinct absolute size difference over the chimpanzees (Figure 10).[7]

But the striking aspect of the figure is that if one adds the various known fossil species of *Homo* and *H. sapiens*, their slope is 1.73. So the relative increase in brain size is a vastly different phenomenon in the *Homo* line than it is in either the Australopithecines or the great apes. The latter are similar to what is found in all other closely related groups of mammals.

In order to understand this remarkable and unique feature of the *Homo* line, one further kind of allometric curve should be considered. This is the relation between brain size and body size during

[7] In other words it has a higher value for b in the allometric equation, or a higher *EQ* in the terminology of Jerison (1973).

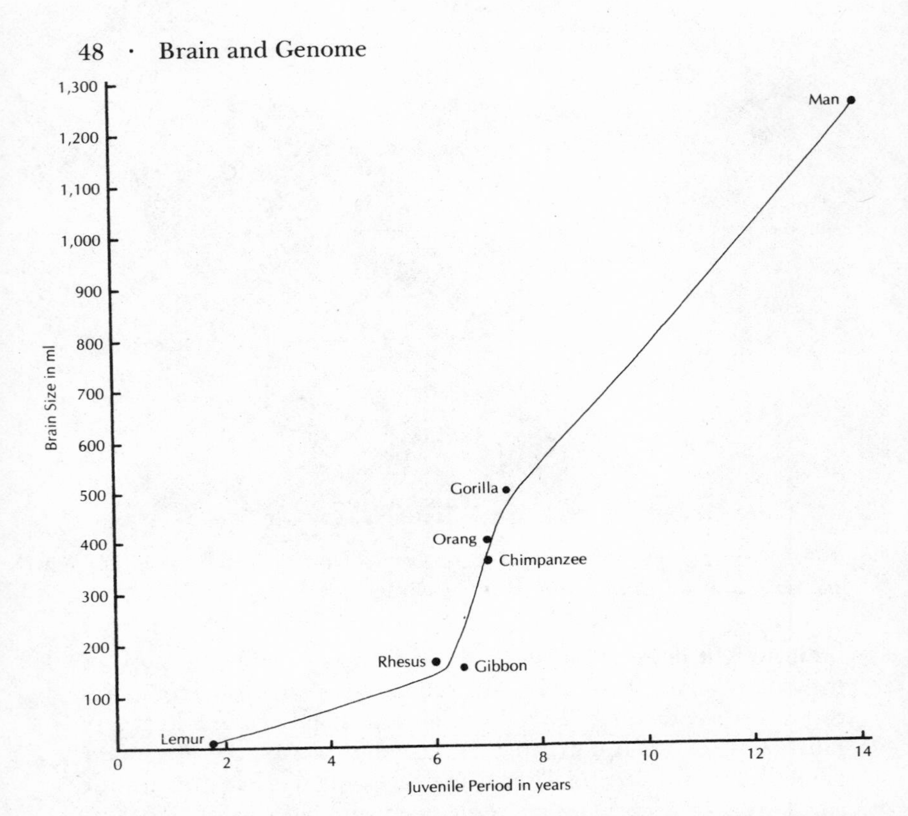

Figure 9. The brain size of various primates and man plotted against the duration of the juvenile period. (Data on the brain size from H. J. Jerison 1973; data on the juvenile period from J. R. Napier and P. H. Napier 1967.)

development. Again this has been examined in detail for man and monkeys and other animals. One of the most extensive studies is that of E. W. Count (1947). He has shown that there are three periods for such developmental allometric growth. In the first, during early development, there is a straight line with a positive slope (1.3) so at foetal stages the brain is growing slightly more rapidly than the rest of the body. This is followed by a transitional period through infancy when growth tapers off. It is during this period that cell division slowly ceases in the brain. The final period is indicated by another straight line that has a slope slightly higher than horizontal. This small size increase is due entirely to an increase in cell size, not cell number. Such curves are shown for man, a chimpanzee, and a rhesus monkey in Figure 11.

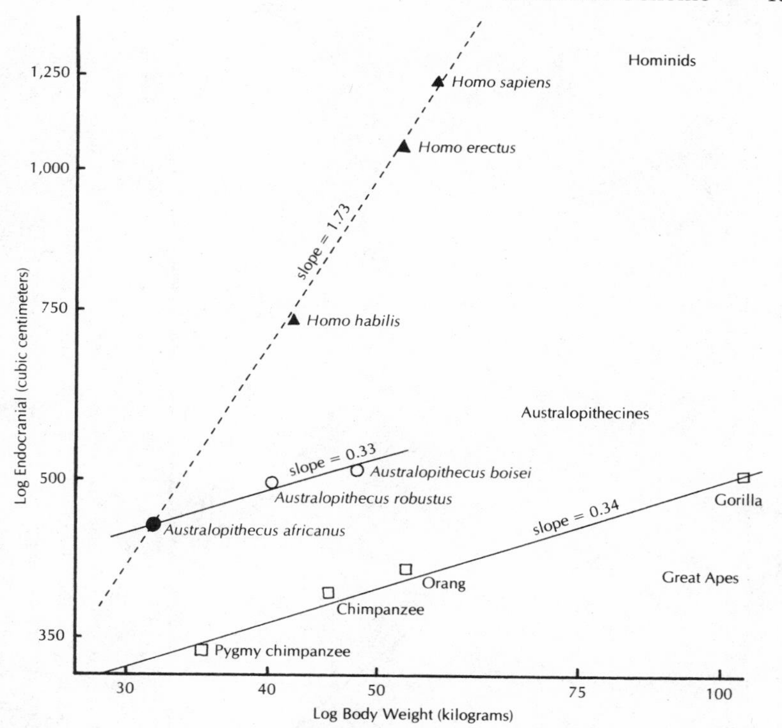

Figure 10. The endocranial volume plotted against the body weight of great apes, Australopithecines, and the *Homo* line on a logarithmic scale. For the fossil forms the body weight is, of course, only an estimate. (Redrawn from D. Pilbeam and S. J. Gould 1974.)

Note that I have also added the points of Figure 10 onto Figure 11 so that it is possible to compare the adult differences and the developmental differences simultaneously. First of all, it is evident that if one compares the ratio of the growth rates in the rhesus monkey, the chimpanzee, and man, they are clearly the same during foetal stages (see Holt et al. 1975). Even the transitional period and final almost horizontal period show the same rates for all three species.

The difference in the final, adult brain size between a rhesus monkey, a chimpanzee, and a gorilla is understood in terms of body weight. The brain of a gorilla is larger than that of a rhesus monkey partly because of the need for more neurons to govern a larger body; but, as indicated earlier, there is the possibility, which was

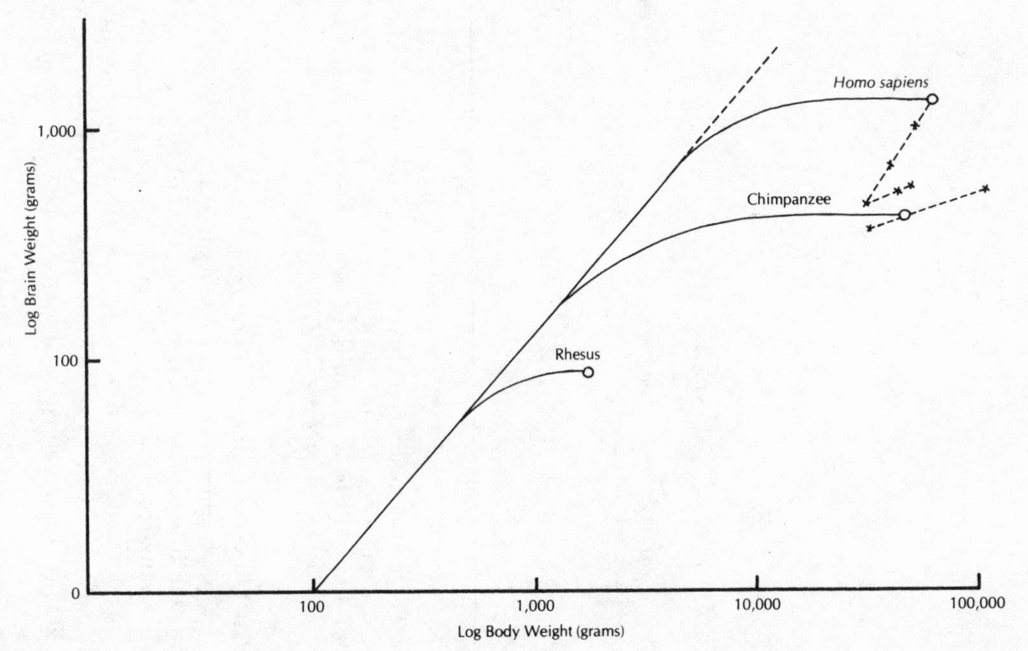

Figure 11. The relation of brain weight to body weight during the development of man, a chimpanzee, and a rhesus monkey. Figure 10 has been included in dotted lines on the right, showing the adult brain-body ratios for the great apes, Australopithecine, and *Homo* lines. (After E. W. Count 1947; D. Pilbeam and S. J. Gould 1974.)

explored by B. Rensch, that brain size increase due to body size increase may give the larger animals some advantages in learning ability and intelligence.

The crucial and important difference in the *Homo* line is that not only does the brain grow relatively longer, but the body stops growing relatively sooner, which is reflected in the fact that the logarithms of mature brain over body size rise to approximately 1.7. Therefore the timing of the brain and body growth is such that they will invariably bear this relation one to the other, while for Australopithecines and great apes, the ratio is nearer 0.33.

If this relatively small change in the development of man, involving simply a difference in the timing of developmental processes, led to man's increased intelligence and elaborate culture, one might well ask what kind of a genetic change could produce such a result. The idea that genes affect rates of processes has been recognized for a long time. To begin, genes code for enzymes that are proteins governing the rates of chemical processes. On the next level of analysis, as Richard Goldschmidt (1938) showed years ago, many genes are known to affect the rates of processes, such as the pigmentation of eyes, or wings, or the rate of growth. Cases are also known in which the genes affect the time when a process, such as growth or pigmentation, starts and stops. How such genes operate and how the sequence and timing of development is controlled is poorly understood, but that such control exists and is one of the main ingredients of normal development is universally accepted.

If one applies this information to *Homo* neoteny, one comes to the conclusion that the genetic difference between *Homo* and the great apes could be very slight indeed: there has been a simple modification of a few genes governing the duration of development of some parts of the organism. As is so often the case, it is more difficult to be certain how selection produced this result. One could imagine that selection pressure for an increase in intelligence favored any genetic change producing an increase in the size of the brain. Such size increase could, theoretically, be achieved a number of ways, one of which would be by neoteny. This method would have the advantage of involving very few gene changes, and as one can see from Figure 11, brain size increase by neoteny did occur. No doubt many other genetic changes affecting brain structure, some of them undoubtedly quite specific, also arose by selection during the early evolution of man.

The idea that few genes are involved in the differences between man and the great apes has received support from the important paper of M-C. King and A. C. Wilson (1975). They have shown that in terms of structural genes (examined by comparison of known proteins, using electrophoresis) there are remarkably few differences between a man and a chimpanzee, far fewer than one would have expected on the basis of the morphological differences. From this they conclude that "a relatively small number of genetic changes in systems controlling the expression of genes may account for major organizational differences between humans and chimpanzees" (1975: 115).

Let me give an imaginary model of how the genes could produce such differences in the timing of the growth of the body and the brain. It is well-known that growth processes in mammals are governed by hormones, most especially the growth hormone of the pituitary gland. The amount of growth is primarily dependent upon two factors: the amount of hormone circulating in the blood and the number of receptor protein molecules for this particular hormone in the cells of the body. There will be little growth if there are either few receptor molecules despite an excess of hormone or if there is a hormone deficiency despite an excess of receptors in the cells. Assume that the genetic change of man's early progenitors was one resulting in an increase in the number of growth hormone receptors specifically in the brain cells during a defined period of development. This means that the brain would continue to grow for a longer interval of time in those species with this genetic change.

Presumably there must be some sort of limit to the increase in relative brain size because, as a number of authors have pointed out, the brain in the hominid line reached its present size level some 300,000 years ago and has not changed since. It is hard to know why this sudden stop occurred, but guesses are easy. For instance, it could be imagined that any further extension of the neotenous development would have been selected against because it produced some deleterious traits. These could be as straightforward as the appearance of imperfections and weaknesses in the physiological integrity of the individual. Any inherited physiological malfunction would be sharply selected against and produce the abrupt cutoff in brain size.

In the discussion so far I have shown the fundamental dichotomy between the transmission of genome information and the transmission of brain information. Furthermore, it is clear from what has been said in this chapter, that while the evolution of the genome has been steady, the more recent evolution of the brain has been relatively sudden and dramatic. Since culture depends completely on the transmission of information by the brain, it is not surprising that with the appearance of man during the course of evolution, culture should shoot up to a great peak, outdistancing anything found in even closely related primates. But before we look at the recent trends of this evolution, let us examine the origin of the brain-genome dichotomy.

CHAPTER 4

The Early Origins of Cultural Evolution

That the difference between the two kinds of evolution resides in a quick, flexible response for cultural evolution and a slow, ponderous response for genetical evolution has been repeatedly stressed. Here I shall look back on early evolutionary events and see where this difference arose. It involves not only separate mechanisms of response to the environment, but the rate of the responses. First we shall isolate the time element and consider that in some detail, and then we can examine the earliest evidence for the dichotomy. Genetical evolution has of course existed as long as there has been life. The pertinent question to ask is where the first rapid, motile response appeared. As we shall see, it is already full-blown in the lowliest of bacteria, and, furthermore, bacteria are known from recent work to provide a splendid system to analyze the molecular basis of the swift response. Consequently, the brain, the whole neuromuscular system, and therefore the capacity for culture, are foreshadowed in the most rudimentary organisms and known to have originated in the earliest of all living forms in the pre-Cambrian period. Bacteria are not capable of culture themselves; they do not have the capacity to teach and learn, or at least these abilities have not yet been detected in them. But, as we shall see, they do have the basic response system.

This is not true of all bacteria; many are nonmotile forms and this immediately raises the interesting question of the selective advantage of motility versus nonmotility, and the reason for the presence of both in all ecosystems. I should I ike to examine this important matter, for it seems to reveal a basic difference in strategies ultimately leading to the brain-genome dichotomy and to culture. In this chapter we shall follow that path from the beginning, always questioning the ecological significance of each stage along the way.

Time

Many authors have stressed that different kinds of biological processes take different amounts of time. Here I shall use a classification of J.B.S. Haldane (1956) that is fairly representative. He considers the following five levels: molecular, physiological, developmental, historical, and evolutionary. This sequence is one of increasing rates. Molecular processes are very rapid and chemical reactions may occur in a range of times from 1×10^{-5} seconds to 1 second. Physiological processes, such as the transmission of an impulse along a nerve or the contraction of a muscle are comparatively slower, and take roughly between 1×10^{-2} seconds to 1 hour. Developmental processes are equivalent to the times of a life cycle, and they vary from 0.50 minutes (a bacterium) to 70 years (a giant sequoia; see Figure 1). Haldane uses the term historical for processes that last a number of generations and may involve millions of individuals, but are of too short a duration to produce the major genetic changes one expects from biological evolution. As an example he cites the spread of the fulmar all over the coast of Britain in the last 100 years. The last category of evolutionary progress is what we have called genetical evolution. It normally takes about 10,000 years to produce a new species, although there are rarer but well-known cases, such as those involving hybridization where new species are formed, or rather begun, in one generation.

He describes these levels as hierarchies: "Each of the processes considered is built up of a very great number of processes quicker than itself. A muscular contraction is the resultant of thousands of millions of molecular transformations. The growth of a limb is the resultant of thousands of millions of cellular divisions, and the acquisition of a skill is the resultant of millions of muscular contractions guided by the nervous system. A historical process is the outcome of millions of lives. An evolutionary process is the resultant of many historical processes" (1956: 387).

It is clear from Haldane's observations that the duration of a process is related to its size: his time hierarchies are also size hierarchies. This is even true within any one level. Small molecules move faster than large molecules; small muscles contract more rapidly than large ones, which A. V. Hill (1950) calls the intrinsic speed of muscle; small animals develop more quickly than large ones (Figure 1 again). In the case of history and evolution the point is even more obvious: the longer the history or the evolution, the more individuals and therefore the greater the total biomass involved.

The relation between size and speed of movement extends beyond living organisms; it is equally true of mechanical devices. The pistons of a small automobile motor move rapidly when compared to the huge cylinders of an ocean liner and a small pendulum swings at a faster rate than a large one. But one can go further than the limited sphere of mechanical devices of different sizes and show that there is a most interesting relation between size and speed of movement if one goes from elementary particles to large stars (Figure 12). From electrons to large molecules, the speed of motion decreases with increasing size, and with larger organisms up through celestial bodies, the speed increases with size.[1] The reason

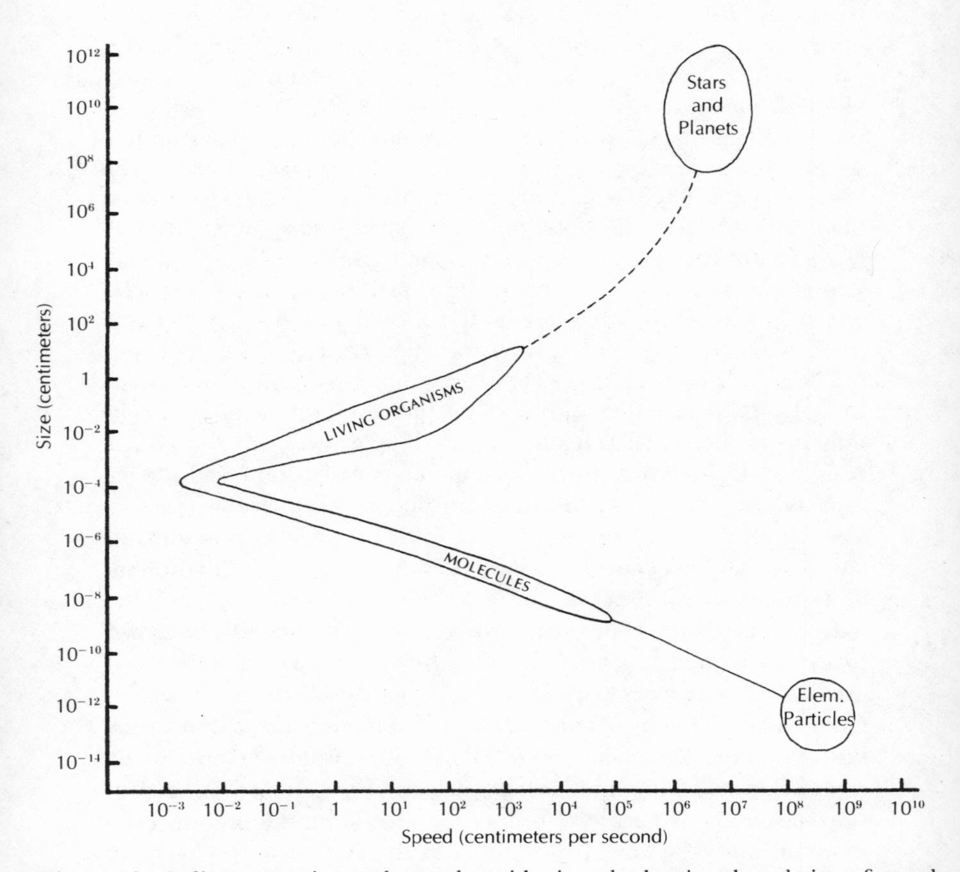

Figure 12. A diagrammatic graph on a logarithmic scale showing the relation of speed of movement to size, from elementary particles to stars.

[1] If one compares the maximum speed the whole organism can achieve, one finds

for this curious relation is not clear to me; it is obviously a problem to be solved by the physicist. However, what is pertinent is the point where the two curves meet. This is the point where a small object is too large to move by physical forces; Brownian motion no longer has the ability to budge the particles for they are too big. Note also that this is the point of minimum size for living organisms. In other words, if it is important for selective reasons for organisms to move, primitive bacteria must devise a motor. It happens they have developed a most efficient and ingenious one, but more on this later.

The kind of motion that interests us particularly occurs when the organism responds to the environment. The fastest response will come when an organism moves away from an undesirable location or toward a desirable one. The slowest response occurs when an organism cannot move. If it is in a favorable location it thrives, but if it is in an unfavorable one it perishes. This involves zero motion.

Various responses require an intermediate amount of time. They are largely developmental examples, although the first one I shall give is only partially so. All multicellular organisms have a hormonal system, and in animals this is in addition to the nervous system. The hormonal response to any environmental change is conspicuously slower than a nervous response. Nerve transmission is rapid; hormones are transported through a vascular system and are correspondingly slow. A phototactic animal sees light, using its eyes or a similar sense organ that transmits the information to the brain. In fractions of a second the brain will tell the limbs to move away or toward light, depending on the animal and the circumstances. In a plant, the light causes the redistribution of auxin in the shoot, and after a number of hours the shoot will grow toward the light. Even in cases where no growth is involved, such as leaves following the rotation of the sun throughout the day, response is slow. The movement of the hormone takes time, and the response by growth or movement of water content of cells is extremely sluggish when compared to muscle movement.[2] In animals there can also be relatively slow responses, such as a rise of adrenalin in the blood stream accompanying sudden anger or fear, but this case is not clear-cut since adrenalin is a neurohormone.

There are numerous instances where an organism will respond

---

that the larger the beast, the faster it moves. This is true for swimming, running, and flying. (See Bonner 1965: 185ff.; Heglund et al. 1974.)

[2] Of course, there are a few exceptional plants, such as the insectivorous Venus' flytrap, whose trap leaves can close on an unwary insect with surprising speed.

to the environment by an appropriate developmental change. A number of such cases in higher organisms have been described by I. I. Schmalhausen (1949). A clear example is the case of a water plant such as arrowhead (*Saggitaria saggittifolia*) where the submerged leaves have a totally different shape from the leaves in air (Figure 13). Another case is the distribution of black pigment on the extremities of the otherwise white Himalayan rabbit. The cold temperature causes the black pigmentation; if the animal is kept in a warm place while it grows, it will remain totally white. In these

Figure 13. Two forms of arrowhead (*Saggitaria saggittifolia*). The terrestrial form (left) and the fully developed aquatic form (right).

cases each leaf or each pigment cell has an alternate choice, and the environment triggers a particular developmental pathway. An even simpler example would be the innumerable fungi that fruit only when the environment shows signs of no longer promoting vegetative growth. The signal can be a drop in humidity, although most often it is a depletion of the food supply. Even spore-forming bacteria can be induced to take the spore developmental pathway by appropriate loss of nutrients in the medium. While these developmental examples are intermediate and give differential responses, it is easy to see they are slow and rigid when compared to the motile responses generated by the nervous system.

Another way to consider the time difference in response is to compare a number of responses to one environmental change in one organism. I shall do this by using a specific example (Figure 14). If man moves suddenly to high altitude (17,000 feet), the first

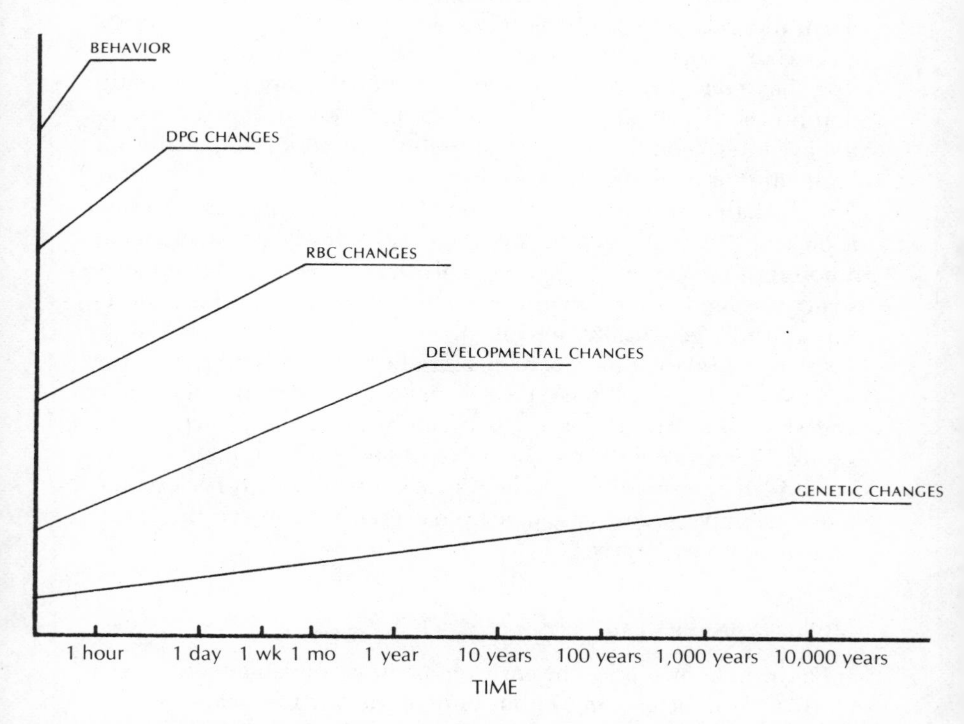

Figure 14. Diagrammatic graph showing the time required for different adaptations to a sudden change in altitude in man.

immediate behavioral response will be to rest and avoid exertion. This fast response is entirely mediated by the nervous system. The two next responses will be physiological. The more rapid, taking some six hours or so, will be an increase in the diphosphoglyceride (DPG) in the blood. This shift in DPG has the good effect of enhancing the release of oxygen in the tissues, but it has the bad effect of causing a decrease in the amount of oxygen that can be picked up in the lung. This, therefore, must be considered an emergency measure that prevents the tissue cells from serious anoxia, but does it at a severe cost. The slower physiological response will be the increase in the number of red blood cells and therefore of hemoglobin. This increase takes a month and the percent of red blood cells (by volume of whole blood) will go from 46 percent (at sea level) to 60 percent (at 17,000 feet). The next reaction to the change in environment would be developmental. There is considerable evidence from populations high in the South American Andes, that to be born and raised at great altitudes will result in an exceptionally large chest, with great lung size. This does not seem to be an inherited character; the rarified atmosphere directly imposes the condition on the developing child. The last and most prolonged evolutionary effect would have to be conjecture. By comparing man with high altitude mammals such as the vicuña and the llama, it is clear that the latter have in some way modified their hemoglobin so that it picks up more oxygen in the lungs and still releases a sufficient amount of oxygen in the tissues. This change seems to be one of a structural modification of the hemoglobin molecule, which therefore must be genetically controlled.

All these adaptations to high altitude beautifully illustrate the wide spectrum of adaptive devices involving different time scales and show that the dichotomy between behavioral and genetic responses has, during the course of evolution, spawned numerous responses of intermediate reaction speed. But if we go back to the most primitive organisms, namely bacteria, we can see the initial basic dichotomy clearly.

Motility in Bacteria

It has been known since the early observations of van Leeuwenhoek in 1676 that bacteria move, but only in the last few years have we had any understanding of how they do this. It had always been imagined that the flagella of bacteria whip back and forth like a snake, bending alternately on one side and then on the other. This view has been reinforced in the last twenty years, because the flagella of protozoa and other higher cells (eukaryotic cells) apparently move in exactly that fashion. They are made up of nine pairs of microtubules that form a cylinder and run the length of the flagellum. Somehow these tubules seem to contract and expand so that the flagellum can wave back and forth. But the flagella of bacteria (a prokaryote) are much smaller and have no such elegant microtubular structure.[3] They are, in fact, one simple thread of a particular protein (flagellin) with no obvious movable parts, so the cause of locomotion is at first sight an impossible problem; and some authors had even decided that the flagella are not involved in locomotion, but just flap passively in the breeze as the cell rushes along, propelled by some other means.

The first glimpse of what is really happening was provided by H. C. Berg and R. A. Anderson (1973; Berg 1975). They made a suggestion that seemed totally absurd from the mechanical point of view: the whole flagellum rotates with respect to the cell body of the bacterium. The motor within the cell can somehow make the flagellum turn at its base, and furthermore it can change its direction and go either clockwise or counter clockwise (Figure 15). These findings were confirmed shortly afterward by some convincing experiments of M. Silverman and M. Simon (1974). In one experiment they attached latex beads to the flagella and showed that the beads spun around the axis of each flagellum. They also tethered the flagellum to a glass slide, using an antibody technique, and then the cell proceeded to rotate wildly around the immobilized flagellum. In cells with more than one flagellum, which is commonly the case among bacteria, all the flagella rotate together at the same speed and direction, acting, therefore, as one (Figure 15).

It has been known for many years that motile bacteria were capable of chemotaxis, that is they would go toward high concentrations of some substances (positive chemotaxis) and away from others (negative chemotaxis). This was first shown by W. Pfeffer in the 1880s who placed a capillary tube with a test substance in the midst of a suspension of bacteria. If the tube contained an attractant, many bacterial cells would enter it. This early observation has been

[3] An eukaryote is an organism whose cells contain a distinct DNA-containing nucleus bounded by a membrane. They may be contrasted with the more primitive prokaryotes whose DNA is not bound within a nuclear membrane. Bacteria and blue-green algae are the main groups of prokaryotes; all other organisms are eukaryotes.

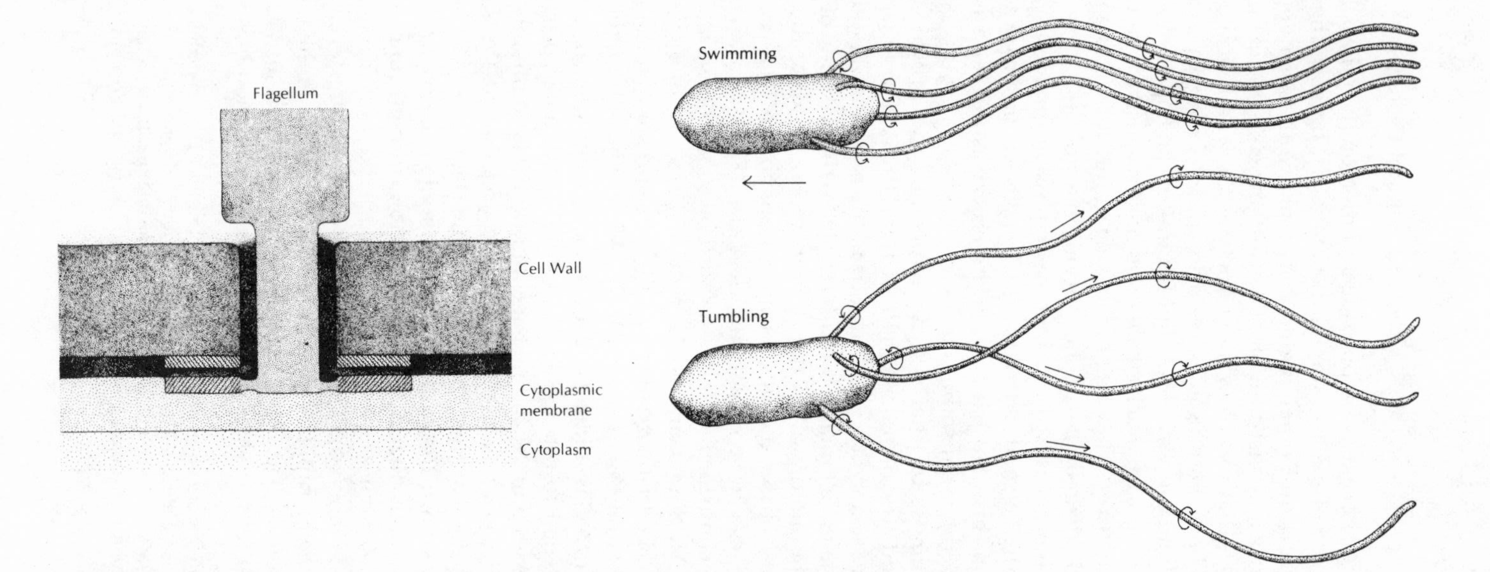

Figure 15. A bacterium (*E. coli*) swimming forward showing the rotation of the flagella (top right), and how a sudden reversal of their rotation causes them to splay out and the cell will tumble (lower right). Above left is shown the simplest kind of rotatory motor of a bacterial flagellum. (After J. Adler 1976 and H. C. Berg 1975.)

beautifully exploited by J. Adler (1976). He first made the assay quantitative and then looked for mutants of various sorts, using the standard analytic procedures of the molecular geneticist. Of particular note is his discovery that a mutant unable to metabolize the sugar galactose still moves up a galactose gradient. This led him and his co-workers to the discovery that specific receptor proteins at the bacterial cell surface combine with the attractant (or repellant) and somehow affect the locomotory apparatus. Twenty such receptor proteins have now been identified for one species of bacterium (*Escherichia coli*), and they are specific for twenty different chemotactic substances (twelve attractants and eight repellants). This is, in essence, the sensory apparatus, and it is thought that it works basically the same way as our own sense of taste or smell.

The next questions to consider are how and what this sensory part of the system communicates to the motor system. We only know the answer to part of this story. H. C. Berg and D. E. Koshland and their respective co-workers have shown how the bacteria behave in gradients of different sorts. Without any gradient an individual bacterium moves in a straight line lasting a second or two and then it "twiddles" or "tumbles" and starts off in another direction, the net result being a random staggering. If such cells are put in a gradient of an attractant, then as long as they move up the gradient, the tumbling is suppressed. The inevitable result is that it will proceed for longer runs in the up-gradient direction. This was shown in some elegant experiments where the cells were in a chamber in which the concentration of a substance could suddenly be shifted. For an attractant, any increase in concentration caused a suppressing of the tumbling; a decrease or no change had no effect on the frequency of tumbling. What actually happens in the cell during a tumble is that the flagella temporarily reverse their direction of rotation. This causes the cell to stop moving, and it becomes reoriented when the flagella go back to their normal direction of motion.

There are many important questions still to be solved. We do not know how the rotor at the base of the flagellum actually moves, and we do not know how the receptor protein can inform the rotor to alter its direction and tumble. These are large questions and the answers may be long in coming. But we do know that despite its small size, it can do some remarkably complex things. Studies have been made where cells are put in two gradients simultaneously: a repellant and an attractant. The cells have specific receptor proteins for both, and apparently they will respond to both simultaneously if the concentrations are properly adjusted. The double gradient will be equivalent to no gradient at all as far as the cell is concerned, and it will move randomly. The two separate bits of external information are somehow integrated and a compromise is reached.

Another very interesting point was made by R. M. McNab and D. E. Koshland (1972) from their experiments in which the concentration was suddenly changed in a small chamber. The minute bacterial cell apparently does not sense that it has more of the attractant molecules on one end as compared to the other, but rather it compares the concentration at two separate times. If at one moment there is an increase in the concentration of an attractant over the next, then tumbling is suppressed. As they point out, this comparison must involve a short term memory, and they suggest a model involving two enzymes that act at different rates producing a lag that would permit comparisons of two readings at two moments, one immediately following the other.

As is plain to see, until recently bacteria have been seriously underestimated as organisms capable of behavior. They can take quite complex readings of their immediate environment and directly translate the information into the appropriate directed movement.

Motile versus Nonmotile Bacteria

Our argument so far has been that bacterial motility is a primitive forerunner of brains and nervous coordination in general. We now want to compare these motile bacteria with the myriad of bacteria that are nonmotile and ask how two similar organisms with such different life strategies can coexist.

Nonmotile bacteria thrive and grow in a favorable environment and die if they are in an unfavorable one. (If they are capable of spore formation, they will avoid destruction by becoming dormant.) Therefore the success or failure of nonmotile bacteria that do not form spores depends totally on whether or not they are transported from one place to another and by chance fall into an agreeable environment. One could immediately argue that this is a substitute for motility. A motile bacterium will swim toward a favorable source of food, while a nonmotile form will get there by chance dispersal. This kind of dispersal does indeed partially substitute for motility, but is a different kind of movement. It is more likely to move the bacterial cells great distances, and be of less significant help over short distances. In this respect chance dispersal will serve as a special mode of movement for both motile and nonmotile species. But over the shorter distances, the percent of the cells failing to find suitable nutrient must be larger among nonmotile forms than among motile ones.

Let us consider the simplest case where a bacterium is nonmotile and has no resistant spore stage. How could this ever have come into being since the lack of motility is a disadvantage? It is well-known that the easiest mutations to obtain in the laboratory are nonmotile strains of motile forms. Assume that the method of dispersal is so favorable that the nonmotile mutants have an equal chance of survival with the motile forms. Then clearly they are not at a competitive disadvantage and the two forms will coexist. It is conceivable that in some circumstances, in a particular location, all motile forms might disappear completely, simply because there is no selective pressure for the retention of continuous movement. But in general, bacterial motility must be advantageous for the trait is retained. A sufficient number of genes is involved in governing the locomotory apparatus so that a single mutation (unless it is a back mutation) is unlikely to turn a nonmotile form into a motile one.[4] But in a motile cell any mutation of a gene connected with motility could cripple any part of the system so that the cell cannot move. In other words, chance dispersal alone can suffice under some circumstances; under others the quick cell movement can confer a distinct selective advantage.

There are some interesting forms that have both motile and nonmotile stages in one life cycle. *Caulobacter* provides a striking example: when it attaches to a surface a stalked individual divides, and one product of the fission will be the parent stalk cell, while the other will be a motile swarmer (Figure 16). The swarmer will eventually settle and repeat the process while the original stalk cell will continue to bud swarmer cells. This life cycle would seem to support clearly the contention that not only is the motile stage adaptive, but the nonmotile or sessile stage is as well.

This raises the question of the advantage in remaining attached or anchored. In this case one presumes the reason is that normally *Caulobacter* exists in an aquatic environment where there are cur-

[4] A back mutation is a second mutation that causes the organism to revert back to the original condition.

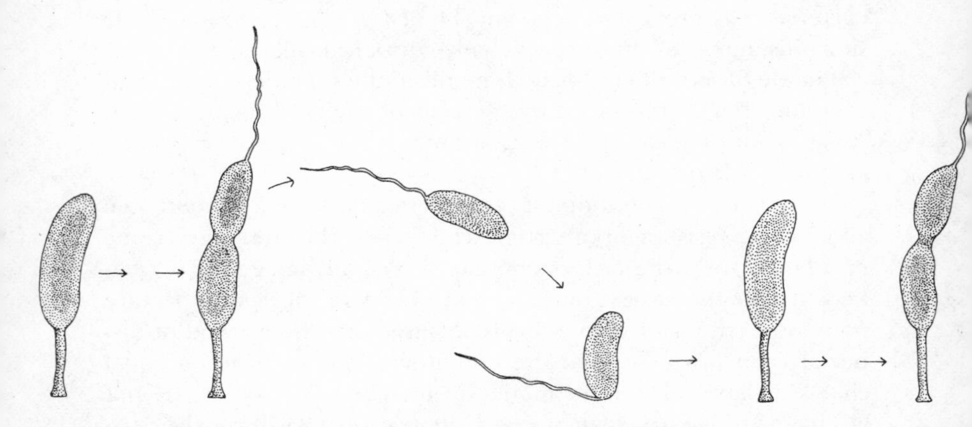

Figure 16. The life cycle of *Caulobacter crescentus*, a stalked bacterium. A stalked individual pinches off a motile cell that ultimately attaches to the substratum, loses its flagellum, and develops a new stalk. A stalk cell will continue to pinch off flagellated cells. (After electron micrographs of B. Terrana and A. Newton 1975, *Developmental Biology* 44: 380-385.)

rents. If there were no means of attachment, all the cells would be carried away. By staying in one place, the current brings the nutrient to the cells. The adaptive advantage of the motile stage would be to colonize new areas.

A more complex state of affairs is found among the Myxobacteria. There the cells divide and feed in a large mass that moves about like a giant amoeba. (In this case it is not known how the cells move for they lack flagella. There are a number of lower plants that can accomplish this mysterious feat.) In this feeding stage the cell mass is phototactic and will move toward light. Besides this advantage, which is gained by their large size, the many cells excrete a massive dose of extracellular digestive enzymes and in this way can eat larger prey than a single cell would be able to manage. After the supply of food is gone, the cells form stalked fruiting bodies with masses of spores or cysts that slide up into the air (Figure 17). Because it is so common among lower forms and has arisen independently so many times in evolution, there is every reason to believe that putting many spores on top of a pedestal jutting out into the air is an effective way to disperse.[5]

The Myxobacteria provide an excellent example of the basic dichotomy, even though they seem advanced when compared to

[5] For a further discussion of this point see Bonner (1970).

Figure 17. The life cycle of a myxobacterium, *Chondromyces crocatus*. A mature fruiting body (lower left) bears cysts, each one of which liberates numerous motile bacterial rods, which swarm into progressively larger groups, ultimately producing a new fruiting body that rises into the air. The fruiting body sequence is at low magnification (a mature one is about 1 mm high), while the swarming sequence is at high magnification (each bacterium is about 0.004 mm long).

more common bacteria, for they have become a social organism where all the cells cooperate to feed and then to disperse. Note that the feeding stage is the motile one, and there is evidence that besides being phototactic, the feeding cell mass can respond to chemical gradients as well. The sessile stage not only is fixed with a stalk, as in *Caulobacter*, but the cells form a resistant cyst that carries them over severe, adverse environmental conditions, thereby making dispersal even more effective. The rapid food-seeking stage and the slow reproductive-dispersal stage are separated in time, and the appearance of each is in response to the fluctuation in the environment. When there is plenty of food, feeding will continue; when there is none, the dispersal mechanism is erected.

Motile versus Nonmotile Higher Forms

To understand more clearly the ecological significance of the co-existence of motile and nonmotile forms, it is best to look directly at sessile higher plants and motile higher animals. Trees and shrubs are nonmotile and many rely on seed dispersal for their continued successful existence. Animals disperse by their own locomotion, be it walking, swimming, or flying. But this power of movement does more than provide a mechanism of dispersal: it also provides them with numerous other advantages. They can forage, chase prey, run away from predators; they can find mates; they can move away from the sun when it is too hot and into it when they are chilled. From arctic and temperate zones they can even fly to tropical zones during inclement seasons when the cold and the lack of food supply become limiting. So, in all, the animals would seem to have enormous advantages over the plants. Why, therefore, are plants not motile?

Of course, many lower plants, especially the smaller algae are motile, but this is not the case for any larger plant. Very likely a fundamental reason has to do with photosynthesis. This form of energy conversion from the sun is a kind of feeding that can suffice without motility; a photosynthetic organism needs only to stay in the sunlight. Dispersal is managed in many ingenious ways, such as the transport of seeds by wind, by animals, and by flowing water. For plants with separate male and female individuals and for plants requiring cross-fertilization even though one plant has both male and female parts of the flower, the male pollen is brought to the female stigma by wind, by insects, and sometimes by birds or bats. So plants have also solved the problem of finding mates. In fact they can effectively perform all their essential living functions: feeding, dispersal, and mating. Natural selection has modified these functions in many specialized ways. If a plant lives in a zone where there is a warm summer and a frozen winter, it can sit out the winter in a dormant state, and grow in the softer months; this is its substitute for migration. The geometry of tree shape and leaf shape has clearly, by selection, become highly efficient so that, according to the size of the plant and its position in a particular habitat, it produces the optimum shape to capture all the sunlight possible.[6]

[6] There are many interesting studies on the strategy of tree and leaf shape. For

While it is true that plants can effectively do all the essential things to live, there is good reason to believe that their nonmotile condition can ultimately be accounted for by the limitations imposed by photosynthesis. A large photosynthetic animal could not acquire enough energy from the sun to do much in the way of moving. Like an automobile running on a small electric battery, it would soon use up its reserves, and it would have to wait for the next sunny day before it could recharge. Or imagine an animal the size of a human being not eating at all and relying entirely on solar radiation for its energy to move about. Not only would it have to carry an enormous sun-catching device, but also it would spend most of its time resting because of insufficient energy. The great exertion of animals can only be maintained by a much greater intake of energy, such as the consumption of large masses of vegetable matter or healthy portions of energy rich meat.[7] Photosynthesis has marvellous advantages, and is of course the source of all life energy, but it also imposes severe restrictions on the organism, and the most obvious of these is the lack of animallike motility.

The direct relation between the abundance of food and the brain has already been pointed out for animals during the course of their evolution. When food is scarce and difficult to find two things will be helpful. One is the possession of cunning so that the hard-to-find food may be trapped. The other is selecting a food, such as meat, that will give much energy in large morsels so that the animal can last until the next meal. This is precisely what photosynthetic plants and sessile or stationary animals have not done: when they are not dormant they need to be eating all the time, yet they still do not get enough energy for active running about.

There must be some other advantage to being nonmotile, for there are many animals that have reverted to a sessile condition. Adult sponges, ascidians, barnacles, and some molluscs such as mussels or oysters are good examples. The larvae are motile and

---

recent discussions, see H. S. Horn (1971, 1975), T. Givnish and G. Vermeij (1976), E. Leigh (1972), and G. Mitchison (1977).

[7] It is possible to make some rough calculations to illustrate this point. If an organism the size of a small mouse (twenty-five grams) were to rely entirely on photosynthesis, it would require a sun-catching device of approximately one square meter. This assumes it can catch full sunlight ten hours of a twenty-four hour day, and the mouse is active only sporadically. One can understand why mice have devised other means of obtaining energy. A human being would require at least a surface of twelve by twelve feet, but this would probably not provide enough energy to carry all the necessary leaves about.

settle on a suitable spot where they metamorphose into a sessile adult, essentially devoid of any significant powers of movement. In these respects they are identical to our previous example of *Caulobacter*. They will close when disturbed (like mimosa, the sensitive plant), but they have not even limited powers for rushing about. In all cases they feed on small particles in the sea water that they sift out, using various clever, filtering devices. Dispersal is achieved in the motile larva stage, and mating by the transfer of sperm through the water, very much as pollen is transported by the wind. They closely imitate plants in many ways, even to having, in ascidians, a system of genetically controlled self-fertilization.[8] As is well known, there is an essentially identical mechanism in high plants. Why did these organisms, whose ancestors were (and larvae are) totally motile, return to this plantlike condition?

It is impossible to know the answer to this question, but one can easily make some reasonable guesses. If the organism is going to filter feed in an area where there is a current, as do bivalves and ascidians, it would be a total waste of energy to rush about, for the current will automatically bring the food to the animal. This assumption is the same made for *Caulobacter*, where it is presumed that the attachment stalk is formed so that the individual cell can feed optimally in a current. (*Caulobacter* of course does not feed on particles, but on substances dissolved in the water.) Therefore these animal forms have achieved, in common with plants, the optimal construction for a particular type of feeding. In these cases it is the best way to take advantage of currents; in the case of higher plants it is the best, or perhaps the only possible way to capture energy from the sun. Natural selection for optimal feeding is then presumed to be the cause of nonmotility in all forms. The idea that loss of motility is easy to achieve by a variety of genetic mutations has already been pointed out for bacteria; here (and in *Caulobacter*) there has been a more elaborate, positive selection for an effective structure to catch dissolved food, particulate food, or the sun's rays.

But by specializing on these kinds of food, these organisms have given up something: they have given up the possibility of ever being the direct precursors of the quick reactions that lead to the development of the brain and culture. What is especially interesting is that the switching entirely to the genome-dominated pathway and away from the brain-dominated pathway has occurred repeatedly

[8] Some ascidians are even known to have cellulose as a structural component.

during evolution, and in each case because it was the best way to make use of a particular source of food energy.

We have seen in this chapter that sometimes the most effective overall life strategy (which includes the capturing of food) is a quick response involving motility, while in other circumstances a stationary existence is optimal. If we look to the early origins of these two strategies, we find them both in bacteria. Therefore the first step toward the capacity of culture was clearly evident in the earliest known living organisms.

CHAPTER 5

The Evolution of Animal Societies

An animal society is a cohesive group of intercommunicating individuals of the same species. We defined culture as the transfer of information by behavioral means, most particularly by the process of teaching and learning, stressing its difference from the genetic transmission of information. This behavioral transmission is communication; at least the words will be used in that sense here. Since both culture and a social grouping are by definition utterly dependent upon communication, it is obvious that the evolution of the social condition will bear a close relation to the evolution of culture. But they are nevertheless quite separate phenomena. For instance, as we shall see, a complex social organization does not necessarily mean elaborate culture. Social existence is a necessary but not sufficient basis for culture.

Communication between living organisms is by no means limited to higher animals. First I shall trace cooperation by communication in broad strokes from primitive organisms up to higher mammals and ask the fundamental questions of why these communication systems have evolved and what is their selective advantage. Then I shall show how the systems of communication have changed during the enormous span of their evolution.

Social Bacteria and Other Primitive Social Forms

The means of communication between individuals in a primitive species may be exceedingly simple, usually involving the use of chemical signals. Furthermore, the communication not only integrates the group, but also may provide some interesting adaptive advantage either in the method of gathering food or in dispersing. The Myxobacteria, which we have already briefly discussed, provide an excellent example (Figure 17). The individual bacterial cells feed in a swarm, in this way conquering large prey that they could not digest as separate cells. Their massive, communal excretion of extracellular enzymes could only be accomplished by cooperation.

It was first pointed out by M. Dworkin (1972) that they operate by the wolf pack principle allowing the group to achieve something denied to separate individuals. They do this even though they are but lowly, bacterial cells. Furthermore, they not only show their cooperation during feeding, but during fruiting body formation as well. They form fruiting bodies that may be as tall as one millimeter and therefore large by comparison to the individual cells. A fruiting body contains many resistant spores or cysts, each one of which can start a new generation when it germinates in a suitable, nutritious environment.

The bacterial or prokaryotic cell is exceedingly primitive; yet it has already acquired a number of remarkable communal properties. Besides the ones mentioned (feeding and fruiting body formation for dispersal), the feeding cells are phototactic and go toward light when they are in a group (M. Aschner and J. Cronin-Kinsh 1970), and they are chemotactic and can orient in a chemical gradient as well. In fact they appear to be able to merge in larger and larger groups mainly because they are chemotactically attracted to one another (see M. Dworkin 1972 for a review). Finally, it is evident that they also have a division of labor during the formation of fruiting bodies: some of the cells become the resistant spores, while others disintegrate and form the large supporting stalk (J. W. Wireman and M. Dworkin 1975). Like social insects, the social bacteria have the equivalent of two castes: the spores are the reproductive caste, and the stalk cells are the equivalent of the workers, for they do not propagate, but altruistically lift the reproductive cells into a more favorable position for dispersal, disintegrating and dying in the process. It is really quite a remarkable series of social feats when one considers that these are brainless prokaryotic cells, the kind of cell arising long before the higher, eukaryotic cells of our bodies and all animals and plants.

It is not surprising that one finds very similar kinds of social organisms among primitive eukaryotes. The various forms of slime molds are made up of amoebae (occasionally flagellated at some stage in their life cycle), and, according to the species, they duplicate all the social behavior we have described for the Myxobacteria. The communal feeding is a conspicuous feature of the large, multinucleate plasmodium of a Myxomycete (Figure 18). The compound fruiting body is fairly universal among the slime molds, and the division of labor is beautifully shown in the cellular slime molds where some cells turn into resistant spores, while others form the

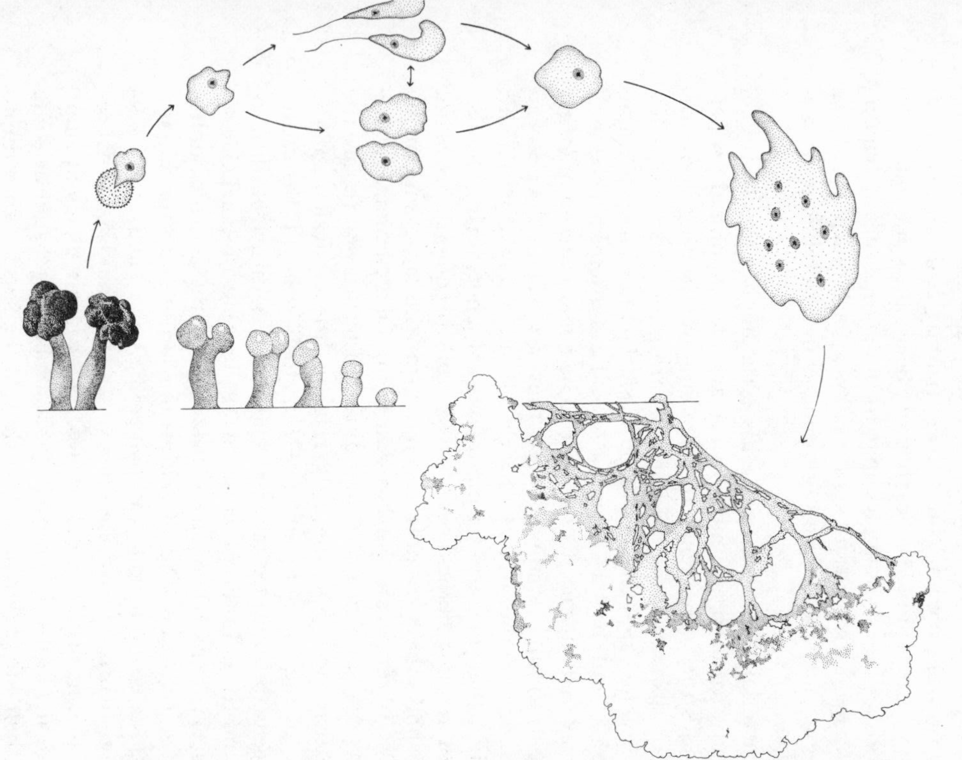

Figure 18. The life cycle of a myxomycete, *Physarum polycephalum*. The minute spore germinates (upper left), giving rise to a cell which, depending upon the environmental conditions, is either an amoeba (dry environment) or a flagellated cell (wet environment). After fertilization the zygote grows into a large multinucleate plasmodium that eventually turns into many spore-bearing fruiting bodies. The lower drawings are at low magnifications; the upper ones of the cells are greatly magnified.

supporting stalk by becoming large and vacuolate, with thick cell walls rather like the pith of a higher plant (Figure 19). Each of these stalk cells dies as it forms, again a perfect case of altruism.

It could be argued that all these examples are no more than the production of multicellular organisms and that this is peripheral to our main concern, which is cooperation between separate individual organisms. There is, however, a continuum. A group of cellular slime mold amoebae that feed as separate individuals and then aggregate by chemotaxis to form a multicellular mass are, at one moment in their life history, a group of separate cells that communicate with one another by chemical signals, and at a later moment they are essentially one, large multicellular individual. If we are to restrict rigidly the use of the word social to individuals of a species that are not physically attached, then these cellular slime molds are social at early stages and simply multicellular at later stages. Furthermore, the Myxobacteria would not be social at all, since they are multicellular at all stages except for the spore stage in some species. Nevertheless they serve as important and useful illustrations, and the question of the degree to which they are social or multicellular becomes something close to a quibble. I include them because they show the fundamental principles of a social existence, and because, at the same time, they are so exceedingly primitive. Furthermore, many of them are exceptional in that they become multicellular by the aggregation of cells, rather than by cell growth.

If one were to concentrate on those organisms that remain separate at all times but yet act cooperatively, then one can find an almost perfect example among the colonial diatoms (Figure 20). The individual cells are small, mobile eukaryotic cells, each encased in a silica shell shaped like a pillbox with a tightfitting lid, and beautifully sculptured. In some of the colonial forms the cells are attached, but in a few species they are quite separate and move rapidly. However, they are confined within a branching tube of extracellular material that they extrude. This tube is attached to the substratum and reaches a size of some centimeters. If one looks at it with the naked eye, it appears to be a delicate, branched seaweed, but if one looks at one of the branches with a microscope, one can see the diatoms within gliding about in a vigorous, haphazard fashion. Unfortunately nothing is known of how they manage to build the very regular, branching tube within which they swim nor what adaptive function it achieves. Hopefully someone will soon undertake to answer these questions.

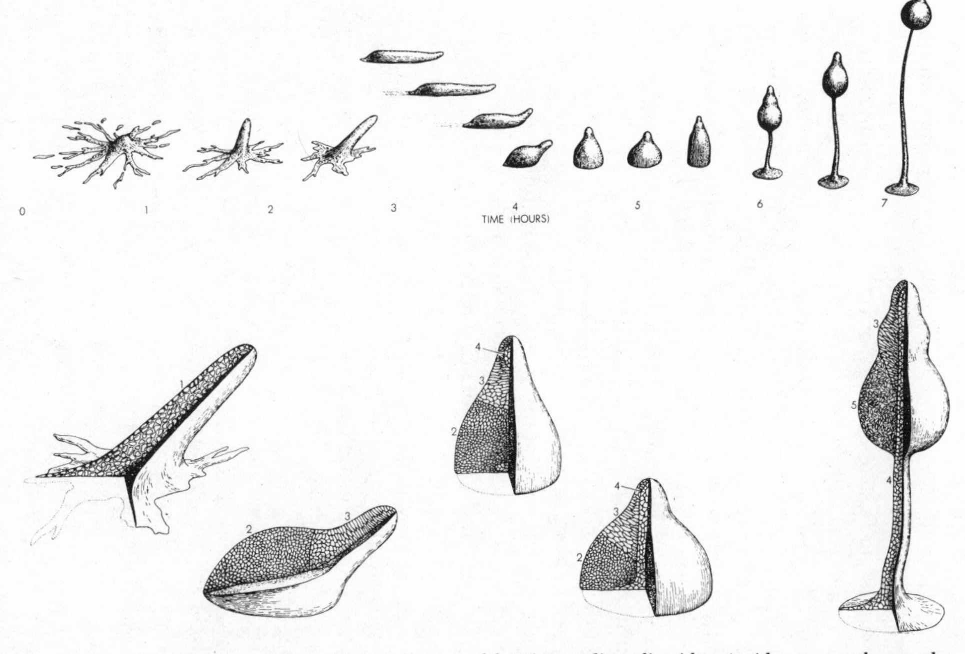

Figure 19. Development of a cellular slime mold (*Dictyostelium discoideum*). Above are shown the aggregation, migration, and culmination in an approximate time scale. Below diagrams are cut-away to show the cellular structure at different stages: (1) undifferentiated cells at the end of aggregation; (2) prespore cells; (3) prestalk cells; (4) mature stalk cells; (5) mature spores. (Drawing by J. L. Howard from J. T. Bonner 1959, *Scientific American*.)

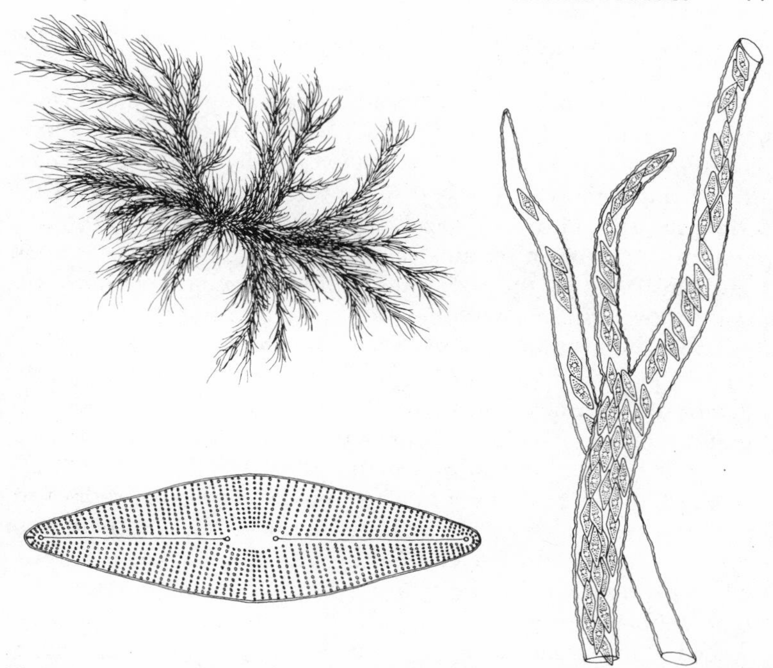

Figure 20. A colonial diatom (*Navicula Grevillei*) showing the entire colony (about 1 cm long): a higher power view of the diatoms wandering about inside the tube they secrete and enlarge, and a high power view of an individual diatom. (After W. Smith 1856, *A Synopsis of the British Diatomaceae, etc.*, London.)

Social Insects

The next level of good examples of social interactions in individual, physically separate organisms does not appear on the evolutionary or phylogenetic scale until one reaches the higher invertebrates. There are many examples where the individuals remain attached to form a compound colony such as the colonial hydroids, including the elaborate siphonophores, and the colonial bryozoa and ascidians. These are of less concern to us because, by being attached, their means of communication is often quite different from those cases where the individuals in a colony are separate. In physically attached colonies the principal means of communication is through cells, often nerve cells; in this respect they are functionally like any single multicellular animal, while this could not be the case if the individuals were separate.

Although it is quite true, as I said in the beginning of this chapter, that culture is dependent upon a social existence (for communication is involved in both), it is not necessarily true that the more advanced and intricate the structure of the society, the greater the propensity for culture. A social existence merely provides the togetherness; the brain permits elaborate culture. This point is perfectly illustrated by the social insects; by all counts they have the most elaborate societies; but, as we shall see in the last chapter of this book, their claim to culture is rudimentary.

In his book on insect societies, E. O. Wilson (1971) discusses the degrees of social behavior in insects and points out that there are three important "qualities" to their being social: (1) there is a cooperation between individuals in the colony to take care of the brood; (2) there is a division of labor connected with reproduction; in other words there are sterile castes in the colony; (3) there is an overlap of at least two generations of individuals in the colony, so that offspring will assist their parents at some moment in their existence.[1] If a colony has all three of these properties, then it is called eusocial, but there are various categories of partially social insects that have only one or two of these qualities. If a species of insect has none, then it is considered solitary.

The highest, eusocial condition is found among the ants, bees, and wasps (Hymenoptera) and the termites. At the same time, however, there are many solitary forms among the Hymenoptera and certainly all the intermediate forms as well. If we apply the same rules to ourselves, we are only weakly social by comparison, although it could be argued that to some degree we manage all three categories. Certainly cooperation of individuals to take care of the offspring is achieved in large families and in day care centers. The amount of division of labor connected with reproduction is flexible and restrained by comparison to sterile castes, although the devoted nineteenth-century nanny in British society came close to filling this role. The best fit is in the overlap of generations; this is one thing we do to a marked degree. But my reason for comparing ant societies to ours is merely so that we have a better understanding of the three properties of insect societies; later we shall examine the differences between insect and mammal societies.

[1] All the information and references given in this section may be found in detail in E. O. Wilson (1971).

Perhaps the most striking aspect of insect societies is the existence of sterile castes, and this matter should be examined in some detail. In many social insects the queen is able to perform all the functions of all the workers. In some ants, for instance, when a queen starts a new colony, she is alone, without help, and she feeds the larvae, not only from her own secretions, but may also go out and gather food. She will build the initial nest with its brood chamber, and she will fight to protect the larvae. As her offspring begin to appear, they will slowly take over many of these duties, and she ceases to perform them. The total duties performed do not increase, but they are now distributed among different castes; usually the smallest individuals will specialize in nursery work, while the largest become the protective soldiers. If all the large workers from a colony are artificially removed, some of the smaller ones will take over the guarding of the nest, and soldiers will help with the care of the larvae if the smallest workers are taken away. So potentially the workers can manage a variety of labors, but under normal circumstances they specialize.

The most developed morphological division of labor is found in both the ants and the termites. In some ants, for instance, not only is there an enormous range in the size of the workers, but the soldiers may differ in shape and have specialized mandibles that are efficient instruments of defense (Figure 21). This is also true in termites, where the soldiers of some species develop poison guns above their reduced mandibles that spray noxious substances at an encroaching enemy (Figure 22).

This division of labor, which involves both behavior and body structure, is so complex and precise that it is important to understand how it comes into being. The evidence is overwhelming (except in one case which I shall discuss presently) that these caste differences are determined by external or environmental factors, that is by pheromones which are hormones passed between separate individuals, or nutrition. There are no caste-determining genes other than for sex determination. The evidence for this has been long in coming and involves the research of many people working on a large number of species of social insects. Note that both hormones and nutrition are chemical influences affecting the development of the insect.[2]

[2] In ants there is evidence also for some nonchemical environmental factors such as winter chilling that might affect caste determination. See E. O. Wilson (1971: 146ff).

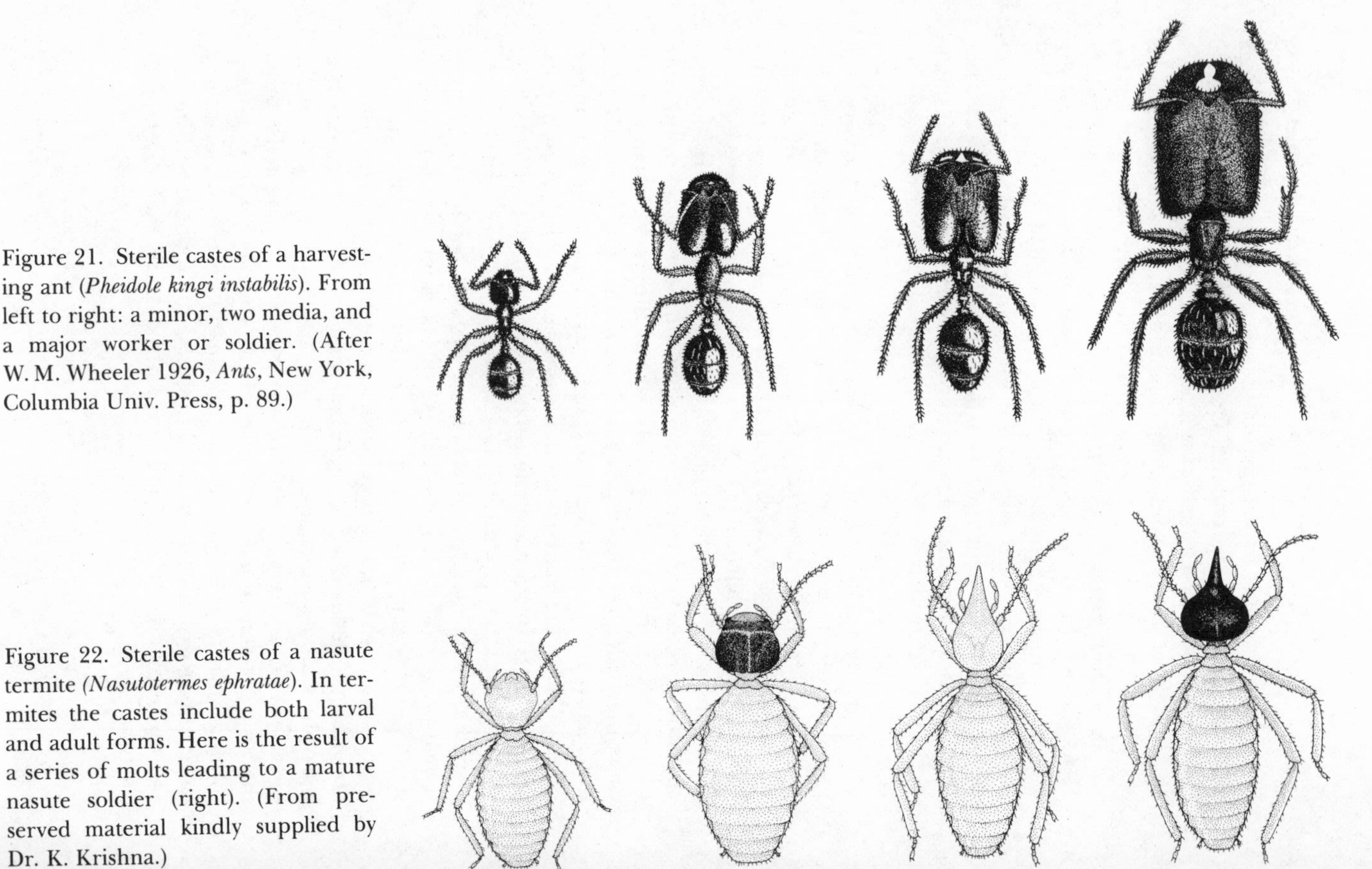

Figure 21. Sterile castes of a harvesting ant (*Pheidole kingi instabilis*). From left to right: a minor, two media, and a major worker or soldier. (After W. M. Wheeler 1926, *Ants*, New York, Columbia Univ. Press, p. 89.)

Figure 22. Sterile castes of a nasute termite (*Nasutotermes ephratae*). In termites the castes include both larval and adult forms. Here is the result of a series of molts leading to a mature nasute soldier (right). (From preserved material kindly supplied by Dr. K. Krishna.)

In some species of wasps and ants it is obvious that the amount of food governs the size of the worker. The best known case of nutritional control of caste is in the honey bee. Special large cells are built in the comb, but the eggs placed in these cells are no different from those of any of the cells that will produce workers. One can interchange the eggs and no matter what egg is in the queen cell, it will produce a queen. The well-known reason is that the workers feed the larva in the queen cell with a nutritionally rich secretion of their pharyngial gland called royal jelly; no larva in a worker cell receives it. The result is a larger bee with active ovaries that becomes a new queen.

Pheromonal control of castes has been shown in many cases, including queen formation in honey bees. A queen in a hive gives off a *queen substance* that inhibits the formation of new queens. There are two known chemicals forming this substance. In its absence (by removing a queen) there will be a rush to build new queen cells by modifying some of the worker cells and rearing new queens. Similar inhibition of reproductives have been demonstrated among termites. It has been shown in some early work of S. F. Light (1943) that if the nymphs of termites are fed a paste made up of soldiers, in the successive molts fewer soldiers will appear, again suggesting control by an inhibiting pheromone, in this case one secreted by the soldiers. Means such as this keep the percent of soldiers remarkably constant in different-sized termite colonies (from 0.75 to 1.00 percent soldiers for colonies from about 6,000 to about 50,000 individuals), as was shown by P. Bodot (1969) for *Cubitermes severus*.

If one considers the evolution of caste formation, it is obvious that the number of castes in any one social insect is related to the size of the colonies. This is particularly evident in wasps where there is a great range of colony sizes, from solitary wasps to rather elaborate and large colonies such as that of the common paper wasp. In some of the simpler forms with small colonies (100 to 400 individuals) one does not see a morphological difference between the worker and the queen; the two castes are distinguished entirely by their behavior. The queen is exceedingly aggressive and shows a very active dominance over the other fertile individuals (Figure 23). She will push and bite her sisters to prevent them from laying eggs into the ready cells, and, should they manage it nevertheless, she will remove their eggs, eat them, and deposit her own. In some other species, which might be considered intermediate on an evolutionary scale, this behavioral dominance system is maintained but

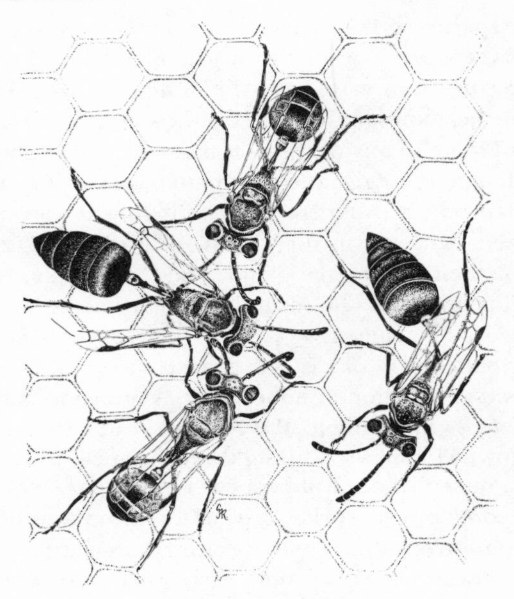

Figure 23. Aggression in a social wasp (*Metapolybia aztecoides*). Two workers dance at a queen (left center), while another queen (right) bends aggressively as they approach. (Drawing by Gerardo Ravassa in M. J. West Eberhard 1978.)

there is a modest size difference between queen and worker. In what is considered the most advanced stage, the aggressiveness has entirely disappeared and the size difference is marked, and there are occasionally differences in body proportions as well. The reason for considering this a more advanced stage is that these colonies are larger, consisting of 300 to 80,000 individuals.

On the other end of the scale, the reproductives (queens) of the stingless bees (Meliponini) have been shown by W. E. Kerr (1950 and the years following) to be at least partially determined directly by genes.[3] That is, one-quarter of the eggs are genetically potential queens and the remaining three-quarters are workers. However, even an individual that is genetically a queen may become a worker if she is deprived of sufficient food during her development. So

[3] See E. O. Wilson (1971) for a review.

again this case appears to be to a large extent dependent upon environmental factors.

I am suggesting that at least among social bees and wasps, which arose numerous times during the course of evolution, a progression has developed from behavioral to chemical control of the division of labor. Let us examine this hypothetical progression. In the first place the genetic involvement has been somewhat oversimplified. The capability of differences in aggression itself could be genetically determined. Furthermore the ability to produce and respond to different chemicals so as to achieve different sizes and shapes must be genetically determined. What is not directly determined by genes is what specific individuals become queens or different kinds of workers; that is determined either by the behavioral dominance hierarchy, or by the balance of pheromones exchanged between individuals, or by variable success in competition for food. So the immediate cause of caste determination is nongenetic. To make an obvious analogy, this is the same as our own bodies where our cells are genetically identical, yet some cells turn into liver cells, some muscle, some nerve, and so forth. The behavioral and chemical control of castes in genetically identical individuals (at least with respect to caste determination) is a developmental phenomenon, in this case occurring during the growth of the whole colony.

Why is it advantageous to have a system of caste control involving behavior or chemical signals; why are they not all genetically determined? The answer is the same as for multicellular development. Both the behavioral and the chemical system provide a way in which one can reach a balance in the division of labor; but more important it is not a rigid set of ratios, but a truly flexible steady state. For instance, if an accident or a war were to eliminate the majority of the soldiers in a termite colony, after a series of molts the ratio of soldiers to workers could be restored. If the castes were genetically determined, this flexibility would be impossible. Even if there were a mechanism whereby the queen could control the number of worker eggs versus soldier eggs, the replacement process would take a much longer time. (Queen Hymenoptera can do this with the sex ratio by permitting or preventing the fertilization of the egg, thus producing a female or a male respectively. The control lies in the sphincter leading away from a small sac that holds the sperm near the oviduct.) By both aggression and pheromones the colony regulates its division of labor in a proportionate fashion, just the way a regulative embryo replaces lost parts as it develops.

From the point of view of natural selection, the advantage of such a flexible system is obvious. What is especially interesting is that there should be two such totally different ways of achieving the same thing: aggression and pheromones. The two are related in the sense that nerves use chemical transmission at their synapses, and insects are particularly well equipped to receive chemical stimuli on their sense receptors. But nevertheless the differences are large. In one it depends upon what one individual does to another; it involves a series of behavioral acts. In the other the effect of the chemical is passive on behavior, but acts directly on the development of the new individuals that will mature in the colony.[4]

The interesting question is to explain the difference between insects and mammals if one assumes that each, in their own way, is adaptive. It is perhaps easiest to see that the more rapid behavioral control system has an advantage by being quick and that it is a method of role determination that suits animals with large brains. But this is inadequate because, as we have just seen, insects also have behavior hierarchies, and they appear to be more primitive than the morphological castes. Therefore we must ask what would be the adaptive advantage for insects in having the slow, developmental caste system.

The answer perhaps is that in insects with distinct morphological castes there is the possibility of combining not only a division of labor of behavioral acts, but of building individuals of great differences in size and proportion that are perfectly suited to carry out those tasks with optimal effectiveness. Defensive and offensive warfare are performed by the largest individuals, while the smallest ones specialize in nursery activities. There are some extreme cases of specialization where, for instance, a soldier ant will have a specially shaped head that perfectly plugs the entrance hole of a nest, and is only pulled back when she receives the right signal from one of her sisters (Figure 24). Another case in harvester ants has recently been described by G. F. Oster and E. O. Wilson (1978) where the so-called soldier turns out not to be a soldier at all, but a worker

[4] There is, in fact, a third method of dividing the labor in social insects. There are examples among bees and ants where a temporal or age difference between workers is used as a means of labor division. For instance, in honey bees it is known that the young workers stay in the hive and attend to housekeeping and brood care activities and as the worker ages she becomes a forager. This is again a system of labor specialization involving the development of the insect. (For details see M. Lindauer 1961; E. O. Wilson 1971.)

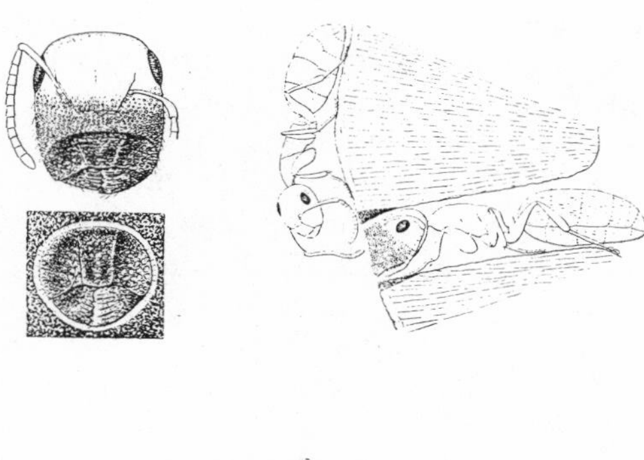

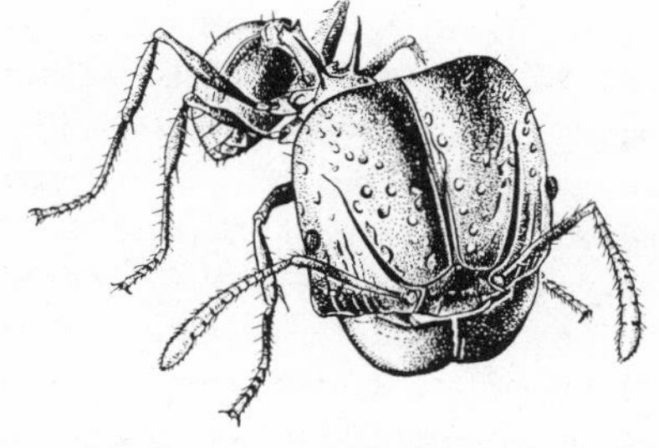

Figure 24. Specialized workers in ants. Above right: nest-guarding soldiers of *Campanotus truncatus*. Their head blocks the nest entrance and will only withdraw when a minor worker signals. Above left: details of the head. (From J. Szabo-Patay, in E. O. Wilson 1971.) Below: a seed crushing or miller worker of *Acanthomymermex notabilis*. (From G. F. Oster and E. O. Wilson 1978.)

that uses its huge mandibles solely to crush big seeds that would otherwise be unusable as food by the rest of the colony (Figure 24). Therefore castes fit specialized adaptive niches and use both morphological as well as behavioral differences to divide the labor.

To a very minor extent vertebrates follow the same scheme. Dominance hierarchies often relate to size, and in some species there may be a marked size difference between the sexes, and often a difference in proportions as well. This sexual division of labor is not necessarily restricted to the division of reproductive functions, although that is paramount; it may also involve other activities, such as differences in methods of feeding.

Finally, we may anticipate our discussion of culture by asking what insect castes have to do with culture. Castes divide the labor and therefore to some degree formalize the behavioral exchange between individuals. This congealing of the behavioral transmission can only mean that the opportunity for producing culture is severely limited. As we shall see later, there is some teaching and learning among social insects, but it is rigid and stereotyped compared to what one finds in mammals. It is the direct result of the comparative simplicity of insect brains.

Social Vertebrates

A social insect colony is an extended family with enormous numbers of offspring, most of them sterile and usually serving one parent or one set of parents. In social vertebrates, and here I shall concentrate my examples among birds and mammals, the social structure takes a considerable variety of forms with the family as only one possibility. As N. Tinbergen (1951) has emphasized so clearly, vertebrates, like insects, interact with one another during the processes of seeking a mate, courtship, and all the rituals associated with bringing the ripe egg and the sperm together. For instance, in the interactions between parents and offspring elaborate signals and response systems between the two ensure the safety and survival of the young. Furthermore all these acts connected with reproduction are under close hormonal control. Not only is sexual activity dependent upon the rise in sex hormones, but as D. Lehrman (1964) showed in such elegant detail for the ring dove, each successive stage in courtship, nest building, brooding, and feeding the young is foreshadowed and governed by hormonal changes that call forth the appropriate innate behavior and do it in the correct sequence.[5] The changes in body chemistry directly affect behavior, and the order of those changes is under rigid internal control.

Of particular interest is the social behavior associated with territorial defense. For many vertebrates, a clearly defined territory for offspring rearing seems to be fundamental. This involves aggressive behavior of a great variety on the part of the male (and sometimes the female as well), usually of a ritual nature, but effective in defending an area that is the sole preserve of a mating pair.

[5] For a current review with the new developments of the subject, see R. Silver (1978).

All of these behavior patterns described so far are fixed. For us it is perhaps of greater importance to consider those social groups that extend beyond the monogamous family, but involve groups of individuals. Of course many types of groups are as stereotyped as the above examples. This is especially true of those directly related to mating such as the formation of harem groups in seals or deer or the formation of leks that are found in a number of species of birds, in some African antelopes, and probably in some species of butterflies. In leks a group of males gather and produce a massive display of courtly splendor that strongly attracts females who come to be fertilized.

The more interesting group, from our point of view, is one in which the individuals remain together either permanently, or at least during the nonmating periods. The female groups with their young of the European red deer or African elephants would be good examples of the latter; wolf packs or numerous monkey groups, such as the howler monkey or the hamadryas baboon would illustrate the former. F. F. Darling (1937) showed that in female or hind groups of red deer there is a dominance hierarchy, the leader being the oldest fertile individual. The organization of the group appears to be centered around a system of protection. This is no better illustrated than by Darling's observation that when a group flees they form a spindle-shaped pattern with the leader in the front and the second dominant individual in the rear (Figure 25). If they pass into a gulley, the last individual will stop and fix the intruder with her eyes while the group disappears. The moment they emerge into sight again, the leader takes up the sentry duty, and the last individual rushes forward to rejoin the group to reform their spindle. In African elephants the oldest female is again dominant, and under attack she will take remarkable risks and often show blind courage. The importance of her role is emphasized by the fact that if she is shot and killed, the group will go into a frenzied disarray for a long period of time. If, on the other hand, another female or a calf is wounded, the whole group, with the old female taking the prime initiative, will try to support and help the stricken individual away from danger.

It is clear from these two examples that their main benefit is that of mutual protection. This, however, is by no means the only advantage of animal groups; in fact often there will be a set of advantages including protection, effective feeding, mating, infant care, and migration. For instance, a wolf pack shows many of these char-

Figure 25. Red deer hind group moving away from danger in an organized fashion. The last hind fixes the intruder while the group escapes into the gulley. She will join them only when the leader (right) can see the intruder.

acters. The North American wolf will form a pack of as many as twenty individuals. A dominant male is the leader, and the other members of the group, including the adolescents, will most likely be closely related. Wolves go through relatively stationary periods when the packs will each have a large territory for hunting, and through periods of extensive migration, especially evident in packs following the migrating caribou. The individuals have elaborate means of communication involving tail waggling, greeting ceremonies, and more aggressive confrontations. Since some domestic dogs are thought to be descended from wolves, we are very familiar with the subtlety of many of these kinds of communication.

A number of remarkable cooperative behavior patterns results. Care of the pups is certainly one. The mother tends the cubs when they are small, and the other wolves bring both her and the cubs food when they return from a hunt. As they mature other members of the pack take turns baby-sitting so that the mother can join the pack in a hunt. The most elaborate cooperative behavior is the hunt itself, which has been described by a number of authors (Murie 1944; Mech 1970; see Wilson 1975 for a review). Numerous instances have been described in which the wolves ambush or outflank a large prey animal, such as a caribou or a mountain sheep (Figure 26). They employ clever tactics involving the use of the ter-

Figure 26. A wolf hunt. Three wolves (right) are trying to stampede Dall sheep toward the two in the gulley (left).

rain and the strategic dispersal of individuals to drive the fleeing prey into the clutches of their pack mates. There is no doubt that a pack can kill animals that an individual wolf could not manage. I could describe very similar behavior patterns in the African wild dog, lion, and spotted hyena, all of which are carnivores.

Turning to primates, let me pick two examples that differ markedly in the degree of evident overt dominance. In the hamadryas baboon of Africa, dominance hierarchies exist in their most intense form. Our knowledge of their social life comes from the careful studies of H. Kummer (1968; see Wilson 1975 for a review) and others. The *group* contains a dominant male with an absolutely unspeakable disposition. He is followed by a harem that may include from one to four or more females, their offspring, and some accessory males. This is the basic social unit, although during some periods of the day, a number of these groups may come together to form a *band* of approximately 40 to 50 individuals. At night when they take to the shelter of trees or cliffs, a number of bands come together to form a *troop* that may consist of between 125 to 750 individuals. The small group seems primarily concerned with mating and care of the young; the band is a feeding unit, and the troop is a unit of protection against the predators of the night. While these animals are totally social, the different functions of their social existence seem to engage different numbers of individuals.

The excessive dominance is shown in the group. The top male actively herds his females together. If one strays, he will rush at her with ferocity. She submits by rushing toward him, and if her submission is not quick or sufficiently enthusiastic, he may slap her or bite her in the neck. He is of course equally fierce with males from other groups or the accessory males within his own. But this aggressive behavior is not confined to the males; the females also have a hierarchy, and the individual highest in their scale is the favorite of the dominant male. If her authority is challenged, she does not hesitate to call in the old male to reinforce her position.

By contrast South and Central American howler monkeys are the gentlest and most democratic of social animals. They live in bands that, in benign periods, may include as many as thirty-five individuals or may be as small as four. Their behavior was first described in a classic study of C. R. Carpenter (1934; see Wilson 1975 for a review). They live in the deep tropical forest, and each band has its territory apparently maintained by deep and rich howling at regular intervals. The males are very much larger than the females and are solely responsible for the formidable howling. Unlike the hamadryas baboon they have relatively few predators. So both from the point of view of danger and steady food supply they live in a relatively easy environment.

The dominance hierarchy is so weak that it is still not clear if there is one or more than one dominant male in a band. All the individuals are remarkably tolerant of one another, and perhaps the main function of any kind of dominance might be in leadership in the direction of the movement of the troop. There seems to be very little concern among the males or females about ownership of the opposite sex during periods of oestrus. The mating is quite indiscriminate, and they all seem willing to await their turn patiently.

It is tempting to consider the difference between these two extreme dominance systems to be related to the abundance of and competition for food and the presence of predators. The life of a baboon on the savannah is much more precarious than forest life. Not only are there leopards and other efficient predators, but food can be scarce during periods of drought. For the howler monkey relatively safe up in the canopy in a region of the world where there is little seasonal change and an ample supply of leaves and fruit, there seems to be small pressure. It is possible that these factors do play a part in governing the dominance system, but there are many exceptions, perhaps too many to make any fast rule. It might be wiser to say simply that a dominance system provides an inner structure to a social group so that it can respond effectively to its needs: mating, rearing offspring, feeding, and protection. The question of whether the system is flamboyant or subtle and reserved is really immaterial so long as it works. Chaos can equally well be reduced by a bawling sergeant in an army platoon or a conductor of a string orchestra. For each the receiving must be attuned to the intensity of the signal, and when this balance is achieved, the group can act effectively. The idea that the same goal can be gained many different ways is an old story, and one we will have further occasion to emphasize.

The important point is that for mammals, as for wasps, aggression and dominance, either blatant or subtle, cannot only give structure to a social group, but can do so by helping to divide the labor. However, nothing is found in mammals comparable to the degrees of difference in the morphology of insect castes. The closest is that in mammals a dominant individual may be the largest or the oldest in the group. The ability to take the top role is the result of external factors such as nutrition and internal ones such as experience; there is no evidence that it is genetically determined and there is much to the contrary. In these respects the division of labor is similar to caste formation in insects. But there is nothing equivalent to the elaborate mandibles of soldier ants or to the spray of soldier termites; all the division of labor is behavioral as it is in the primitive wasps. As I have already indicated, the possible reason for this is related to the increase in brain size. By having more elaborate behavior, mammals can perform a large variety of tasks and do so rapidly to accommodate immediate and sudden needs.

The Adaptive Advantage of Being Social

The cornerstone of modern sociobiology is the concept of kin selection and the role of altruism we have already discussed in Chapter Two. There it was argued that one of the reasons animals became social was to preserve their genes. Closely related organisms have genes in common, and by cooperation they increase their chances of preserving those genes for subsequent generations. This theory has been used by W. D. Hamilton (1964) to explain why a social existence arose independently twelve or more times among the Hymenoptera, but only once among the termites. In the Hymenoptera, because of the fact that the males are haploids and the females diploids, the sister workers are more closely related to one another than in other animals in which both sexes are diploid. Hamilton showed that on the average Hymenopteran sisters share three-quarters of their genes.[6] Termites, on the other hand, are totally

[6] The reason for this is because the males are haploid. As a result, all the sperm of a male are genetically identical and therefore the probability of a gene being shared by the offspring of one father equals 1. But the queen is diploid and her eggs, through meiosis which involves chromosome reshuffling or recombination, become haploid. (See p. 15.) Therefore her eggs are not identical and the probability of a gene being shared by the offspring of a queen equals 1/2. This can be put into a simple equation: for a worker, half of her genes will come from her father ($1/2 \times 1$) and half from her mother ($1/2 \times 1/2$). If these are now added,

$$(1/2 \times 1) + (1/2 \times 1/2) = 3/4$$

On the other hand if both parents are diploid, then

$$(1/2 \times 1/2) + (1/2 \times 1/2) = 1/2$$

Therefore, sisters in haplodiploid social insects will be more closely related than the siblings of any diploid organism.

diploid organisms, and therefore siblings share an average of one-half of their genes with each other. His argument is very simple: togetherness is especially advantageous for gene preservation in Hymenopteran sister societies, and therefore the social condition has arisen independently on numerous occasions. The survival machine has been made more effective, by forming a society. Even though the genetic advantage is less in normal diploid organisms such as termites, wolves, monkeys, and ourselves, it is nevertheless tangible and relatedness could be a factor in cooperation for all organisms. As we said earlier, it is an even greater benefit to social bacteria and social amoebae (slime molds) because in these the genes are identical in all cells, as they are in multicellular organisms. In sum, there are advantages to a survival machine that involves the cooperation of related individuals, and this may be one of the reasons so many survival machines are social.

In fact, two kinds of adaptive or selective forces bring animals together into a society. One is kin selection, which we have just discussed. It is based simply on the notion that for purely genetic reasons cooperation between closely related individuals will come under positive selection pressure. The other selective force promoting cooperation and therefore social grouping are those cases where, by communal activity, certain tasks can be performed by a group that would be impossible to achieve as separate individuals. In contrast to *kin selection*, for convenience let us call this *communal task selection*.

Any task could be one either performed more effectively by a group of individuals or one best accomplished by a solitary animal. Let me give an example of each. As I have just shown, wolves tend to hunt in packs, and there is considerable evidence that by doing so they can kill large animals giving them an appreciable advantage in survival and reproduction. This would be a case of communal task selection. On the other hand the Indian tiger and the South American jaguar hunt alone, and one assumes they do this to accomplish more effectively their prey-catching task. They hunt in the dense jungle where stealth is needed to catch their prey unawares. If they were in a pack perhaps they would have more difficulty stalking their victims unobserved. Were this so, the task would clearly not benefit from communal action. Therefore cooperation in a social group could either be the result of kin selection, or communal task selection (as in the wolf), or both. This means that in some social groups there might be no selection for a communally performed task, and the entire selection pressure maintaining the integrity of the group would be kin selection. Finally, there are many cases of solitary animals such as tigers and jaguars where even kin selection is insufficient to counteract the positive selection for noncooperative hunting.

Let us now move to a survey of communal task selection in different groups of organisms. To do this let us see what other kinds of tasks besides catching prey could be advantageous to a group. In each example given below I must remind the reader that we are almost always guessing at the adaptive advantages and usually have no clue as to whether or not our guesses are correct. All we can say is that they are reasonable possibilities. Let us begin with our most primitive examples of social organisms. I have already pointed out that in social bacteria, social amoebae, and many fungi, togetherness is apparently important for more effective dispersal. The compound fruiting body that is lifted into the air must permit a greater spreading of the spores and thereby produces an increased chance of finding a suitable environment for growth and perpetuation. In these microorganisms the food supply may be exceedingly patchy, and weather changes a constant threat. The wider the spores can be spread, the greater the possibility that further propagation will ensue.

As is abundantly clear from the examples already given, numerous organisms can be social during one part of their life history and independent or solitary in another. In the cellular slime molds, the amoebae feed as solitary individuals, and in some species there is evidence for a mechanism of negative chemotaxis that makes each amoeba move as far away from its neighbors as possible (Bonner 1977). However, for all sexual forms social communication is essential during the mating season. In the case of the tiger, no matter how solitary an individual may be, it can hardly serve as a survival machine if it never mates. Furthermore higher vertebrates require a period of infant care, and at least one of the parents must perform those duties.

Another major reason for a social existence is food gathering. There are many excellent examples where the food taken can only be managed by a group of individuals. We have discussed cases from enzyme production in the Myxobacteria to cooperative hunting in wolf packs. There are many others.[7] One of the most inter-

[7] Chapter Three in E. O. Wilson's (1975) book provides a whole series of examples of benefits derived from cooperative feeding and all the other adaptive advantages summarized in the remainder of this section.

esting ones appears among colonial birds. H. S. Horn (1968) has shown on a purely theoretical basis that if birds were seeking food distributed in patches, that is, in widely separated locations and appearing at different times, and if the birds were bringing that food to centrally located nests, then feeding in flocks would be a more efficient and effective strategy than if the feeding birds were evenly distributed. In some species of birds these flocking advantages are seasonal, while in others they exist all year around; it depends upon the variability of the food supply. But generally the collective food gathering will be a major part of the existence of the animal and, therefore, an extremely potent force in producing a social existence as an adaptation.

Protection is an equally important factor. This was stressed for herds of red deer hinds, female elephants, and troops of hamadryas baboons. Top predators, such as wolves and lions are hardly in need of protection, but all those social groups that are not on top of the prey-predator pyramid are likely to find protective benefits of some sort. Sentinel mechanisms are one of the better known examples. In a large group there will be that many more eyes, noses, and ears to detect danger. And should danger come, by massing in various ways, a group can intimidate or confuse an attacker. The clumping of starlings in the presence of a hawk is a classic case. Tinbergen (1951) pointed out that a peregrine falcon travels very fast when it attacks and is fragile except for its talons, so that pouncing into a tight group of starlings is sufficiently risky to be a deterrent.

There are also those interesting cases where, because of the bunching of the number of prey in space and time, the predator can harvest only a small percent of the population. Here there is no individual protection except in a statistical sense. In his discussion of protection gains in social groups E. O. Wilson (1975) cites the work of G. Neuweiler on the large fruit bats (or flying foxes) of Australia. While roosting, the dominant individuals take the tops of the trees, and the subordinate animals have the lower branches (Figure 27). The latter will more likely be attacked by predators, and the commotion sets off an alarm for all the other more secure individuals. So there is a spectrum of conditions from those in which individuals are sacrificed at random (but the total number is kept small by the social habits of the species) to those where the society is structured so that only the members on the low end of the dominance totem pole are sacrificed. Therefore, when considering

Figure 27. Flying foxes resting. (After a photograph of B. Dales.) Their distribution in a tree is also shown above. (After a photograph of G. Neuweiler 1969, *Zeitschrift für Tierpsychologie* 26: 166-199.)

protection as an advantage for social grouping, one must ultimately distinguish between simple, temporary aggregations, iike the bunching of starlings in the presence of a hawk, and the genuine social structures shown by red deer hinds and female elephant groups. We are more interested in the latter, but there is a continuum between the two extremes.

One final category of selective advantages of group behavior may be found in the modification of the environment. Beavers, for instance, cooperate in the building of dams that benefit both food gathering and protection. These dams may involve contributions from numerous families who together build very large dams beneficial to all the individuals involved. Even better cases are found in the construction of nests among social insects. The huge termitaria of some of the more advanced tropical termites are extraordinary complex structures where the inside temperature and humidity control is an architectural triumph. Even honey bees, whose hive construction is simple since they use some cavity or man-made hive for the outside, have amazing temperature control; they can keep their hive within ± 0.5°C of 35° when the external temperatures are either extremely cold or hot. In these and in many other cases the initial advantages of togetherness could conceivably be achieved by making the self-made environment a buffer against the vagaries of the conditions of the outside world; and this, in turn, could have led to more elaborate social interactions.

There are, no doubt, other adaptive advantages to grouping. For instance, Wilson (1975) cites more effective competition as an additional possibility. All adaptations seem to help, under particular circumstances, some rather basic biological functions such as mating, food gathering, predator escaping, or environmental control. We said in the beginning of this chapter that the forces molding animal societies were parallel but by no means identical with those molding culture. Therefore when we come to the evolution of culture in the last chapter we shall compare the adaptive advantages of culture with those of social groupings. As we shall see, despite some overlap, the two are very different.

The Evolution of Communication Systems

Earlier it was pointed out that communication was the basis of both a social existence and of culture. Here we shall examine the methods of communication at different levels of social organization and show that one of the biggest changes, as one reaches the higher vertebrates, is the subtlety and range of communication. This is because (with numerous exceptions) invertebrates tend to emphasize chemical signals, while vertebrates communicate to a greater degree by auditory and visual signals. As we shall see, the latter two permit an infinitely greater repertoire.

If we look again briefly at the social microorganisms, we find that their social existence is governed entirely by pheromones. For instance, in the cellular slime molds (Figure 28) the germination of the spores is regulated by a substance given off by the spores so that the percent germination is high when the spores are dilute and low when they are concentrated. Since only one or a few amoebae are needed to start a new generation, it might be advantageous to keep spores from germinating when they are numerous. The amoebae then feed, and as previously mentioned, in some species a substance is given off by the cells to keep them equidistant for optimally effective grazing on the bacteria that are their food. After the food supply is consumed, they give off an attractant (acrasin) that brings the amoebae into central collection points. The multicellular masses that result are presumably further governed in their development by a series of chemical substances, but at the moment these processes are poorly understood. However, it is known that the fruiting bodies are nonrandomly placed with respect to one another, and this is, at least in part, due to an inhibitor given off by the collecting mass of cells that prevents any other masses from collecting in its area or territory. Finally, as the fruiting bodies rise up into the air, they give off a volatile pheromone, that is, a repellant. As a result, the spore masses are evenly spaced and presumably in that way optimally placed for spore dispersal. These are soil and humus organisms whose dispersal is dependent upon some small animal crawling past them and picking up the spores by capillarity.[8]

If we take a gigantic leap to social insects, it is striking that the use of chemical signals continues to be the major characteristic of their communication system. The matter is admirably reviewed by E. O. Wilson (1971) who has made a number of original contributions himself. Each of an extraordinarily large number of glands (especially in ants) secretes one or more substances with a specific effect. There are alarm substances that serve as warning signals for the colony, assembly substances that cause an aggregation of ants

[8] For a detailed review and the evidence for these points see Bonner (1973).

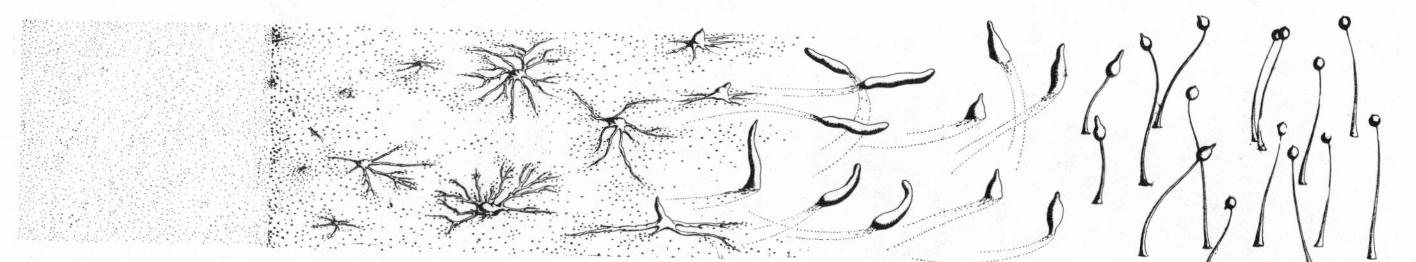

Figure 28. The life cycle of a cellular slime mold (*Dictyostelium discoideum*) from the feeding stage (left), through aggregation, migration and the final fruiting (right). (Drawing of Patricia Collins from J. T. Bonner 1969, *Scientific American.*)

toward food or toward an enemy, and recruitment substances when some cooperative behavior is required such as nest construction or migration. Recruitment substances may involve the deposit of odor trails of various sorts leading the nest mates to food or the nest. Chemical senses are also used in nest recognition and the recognition of nest mates. An intruder with the wrong odor will be unceremoniously evicted. According to Wilson, there is even evidence that members of a colony are able to distinguish between castes; in some cases in small colonies individuals are recognized, in large colonies new and old queens. Most of the kinds of communication described above involve one individual's emission of a substance that another senses by chemoreception. One of the most striking characteristics of social insects is that they are forever licking one another and exchanging liquid food. This supplies enormous opportunities not only for the exchange of nutrients, but also for the exchange of pheromones. If, for instance, a queen termite produces an inhibitory pheromone that prevents female nymphs from molting into what is called a secondary reproductive, the pheromone is passed throughout the colony by mutual licking and food exchange. Such an effect is somewhat different from that of an alarm substance that is rapid and immediate; the former is a developmental response that occurs over a longer period of time. As M. Lüscher (1961) has shown, the inhibition pheromone somehow affects the balance of internal hormones within the developing worker so that sexual maturity is prevented.

A number of social insects use sound vibrations or tactile stimuli for communication signals. For instance, some ant species can stridulate, that is rub two parts of their body together (like a cricket) and make a small chirp used as a cry for help by an individual: other ants will come running toward the distressed chirper. In wasp nests of certain species the larvae in the cells will become active and beg for food at the sound of the buzzing of a flying adult. Tactile communication reaches its highest form in honey bees when a dancing scout bee gives the other foraging bees information about distance and direction of the source of nectar. There will be more about the details of this later, but clearly both tactile and possibly auditory signals play a role. Of a simpler nature, food regurgitation is evoked when one individual strokes another in a particular way, and the stroked individual will part with some food. (Some nest parasites of ants and termites have acquired the trick and can fool their hosts into sharing their food.)

Visual cues for communication are less common among insects. In some species of ants with well-developed eyes, the frenzied activity of one worker will excite another, and the message may be transmitted largely by sight. Again the honey bee uses sight to detect the position of the sun with respect to a source of nectar and can convey this information to bees in the hive by dancing. The hive bees receive the information using their tactile sense, and then go out, translate the information back into a visual form, and fly in the correct direction with respect to the sun. But despite these examples, vision, hearing, and feeling play a relatively smaller role than chemical communication in the life of social insects.

The reverse is the case in vertebrates. This does not mean that they have abandoned the use of chemical senses, but rather that auditory and visual signals have been increased in a variety of interesting ways. Let us begin by discussing chemical communication in vertebrates. While the number of glands secreting signalling substances for any vertebrate is small compared to insects, those gland secretions are very much a part of social behavior. One of the most obvious examples is home range marking in mammals. This may involve urine alone (for example in wolves), faeces mixed with secretions from anal glands (for example castoreum in the beaver), or gland secretions from various parts of the body (for example from near the eye in some antelopes or the musk gland found on the abdomen of the musk deer) that are placed in strategic spots to delimit the home range. There is also increasing evidence that volatile pheromones affect sexual behavior, such as the attraction of dogs to bitches in heat (oestrus) or the synchrony of the menstrual period in women rooming together.[9]

What is especially remarkable about mammalian chemical senses is not the number of substances emitted, but the great sensitivity of the chemoreceptors in the nose. Not only can most wild mammals smell a predator many miles away (provided the wind is blowing the odor to them), but they can recognize their own species and even identify individuals such as their offspring among a mass of similar young. A most striking demonstration of this fact is shown in some experiments run by H. Kalmus (1955) using some of the best police dogs of the London police force. He engaged the help of a number of individuals unknown to a dog, some of them related

[9] This is the work of McClintock (1971). The idea that the synchrony is achieved through the sense of smell is only conjecture; the actual mechanism is not known.

to each other and others not. In one kind of experiment he had a dog retrieve one of a set of handkerchiefs, each of which had been under the armpit of a volunteer. Under these circumstances a dog could distinguish between siblings and any other members of their family by picking up the correct handkerchief, but could not distinguish between identical twins: they smelled the same. However, in a retrieving experiment (using a similar handkerchief as a cue) in which one of the twins and another individual criss-crossed over a field, the dog could distinguish not only a twin from some other member of its family, but also could distinguish between the two identical twins. Therefore, depending on the circumstances, a well-trained dog can distinguish individuals, although the closer the genetic relation of those individuals, the more difficult is the task for the dog. In the introduction to H. Kalmus' paper he gives an interesting anecdote from John W. MacArthur who was a professor of genetics at Toronto, which apparently stimulated Kalmus to do these experiments: "In a prospector's camp in Northern Ontario was a nearly blind great dane, Silva. She was not a friendly dog, but she had a passion for one prospector and fawned on him delightedly whenever he came to camp.

"One day a stranger appeared for the first time and to the surprise of those present, Silva greeted him with great affection. Upon enquiry by the camp crew it was found that the stranger was the identical twin of the particular prospector, whom Silva had such a passion for, and that he had never been there before nor seen the dog before" (1955: 25). It is clear that if the reception system is sufficiently delicate, it can obtain information from the outside world of which we have small understanding because of our own insensitivity. Perhaps a great wine connoisseur or teataster is as close as we can manage.

Despite the subtlety of chemoreception, the number of chemical social signals given off by vertebrates are few. Therefore, beyond the highly discriminating abilities to recognize food, predators, or individuals of one's own species and to attract members of the opposite sex, chemical cues play a relatively minor role in social communication in vertebrates. In some groups of vertebrates, such as birds (with the exception of some vultures, woodcocks, and perhaps a few others), the sense of smell is even more atrophied than our own, so that the high-powered chemical receptors we have been describing are hardly universal.

The role of vision in social communication of vertebrates is considerably greater than that found among social insects. The most obvious reason for this is the perfection of the vertebrate eye by comparison to the compound eye of insects. The vertebrate eye is a marvelous invention, rivalled only by the eyes of some cephalopod molluscs. Obviously the marvel comes not just from the camera-like properties of the eye itself, but from all the miraculous ways in which the image is recorded by the cells containing the visual pigments and the ways in which this information is sorted out first in the retinal interneurons and finally in the visual part of the brain. The social uses of vision by vertebrates are extensive. In the first place recognition of fellow species and individuals is possible. The great sexual dimorphism in plumage found in many species of birds testifies to the use of visual cues in sexual activity. Furthermore, during courtship there is often an elaborate strutting and feather display of the male and, less frequently, the female. Not only does the color or shape stimulate, but also the corresponding movements recorded visually by the courting pair.

Vision is involved too in many aspects of offspring-parent relations. For instance, in ducks or geese the young follow what they saw during a critical period after hatching, a process known as imprinting. Visual cues are also used in food exchange between parent and offspring. The offspring begs when it sees a parent; when it sees the begging the parent responds by delivering food.

Among mammals much is made of individuals in a group watching one another and reading, from gestures and facial expressions, the mood of their associates. The fact that mammals are sensitive in this way, even to other species, is something any pet dog can teach us. They watch our gestures and movements so that, even without sounds, they know when we are cross with them or when it is time for a walk. They associate expression and posture with the cause they have pieced together. On the other hand, if they feel it is time for a walk, they will try various ways of making their idea obvious. In primates, especially the chimpanzee, a large variety of facial expression is full of meaning for the other members of a community. The ability of chimpanzees to use highly complex visual cues is well known from the pioneer work of R. A. and B. T. Gardner (1969) on Washoe, a young chimpanzee they raised to understand the American deaf and dumb language. Not only are chimpanzees able to have working vocabularies of more than one hundred words in sign language, but they also can master some simple grammar.[10] This seems to reinforce the idea that in nature they use vision as a major means to communicate, and they can have a subtle understanding of gesture and grimace. This would fit in with the fact that their vocal repertoire is limited, at least when compared to that of man.

Because of human language, we have come to think of auditory signals as playing a dominant role in social communications. It is important for many vertebrates, although often we cannot understand the language. A few species of fish and virtually all dolphins and whales are active noise makers, and since the brains of the latter two are so large, there may be quite elaborate sound communication that we cannot yet decipher. Birds use song as a means of courtship, territory defense, and recognition. One of the more amazing examples of the latter is displayed by the common murre, a sea bird that lays eggs in a row along a ledge on barren rock islands off northern coasts. It has been shown by B. Tschanz that the young birds still in the eggs learn their parents' call a few days before they hatch, and in this way the chicks recognize their parents right from the beginning of their perilous and competitive existence on the edge of a cliff.[11]

Besides recognition, signals for danger, attraction of mates, and food discovery are all announced by the emission of sounds in the majority of birds and mammals. It is essentially the same use as that made of visual signals, although sound signals have one advantage: they can operate under conditions where the animals cannot see one another, that is, in the night or in the dense forest.[12]

Let us now compare social insects and higher vertebrates (birds and mammals). As I have emphasized, visual and auditory signals are more prevalent in vertebrates, although chemical signals are at least present in most groups. What advantages do visual and sound signals have over chemical signals, other than providing a total increase in the number of modes of possible signals?

[10] Numerous other workers have extended this work with other chimpanzees using the deaf and dumb language, and further insight has also been gained by the use of two other methods: colored plastic symbols for words by A. J. and D. Premack and a computer by D. Rumbaugh, T. V. Gill, and E. von Glaserfeld. For a good review of the field, with a useful bibliography, see R. S. Fouts and R. L. Rigby (1977).

[11] For a discussion of Tschanz's work in English, see Jellis (1977).

[12] I have often wondered if there is a converse. For instance, do birds nest by a thunderous waterfall where they cannot hear song and rely more heavily on visual signals, or do they simply choose some other place to nest?

One possible advantage is that the variety of visual and sound cues that one animal can give another is greater than the number of chemical cues. The repertoire of chemical emissions of even so elaborate a chemical factory as a social ant is limited compared to the gestures and grimaces or the vocalizations of a primate. An ant has to have glands to make the substances, and control consists of how much of each substance is emitted and in what combinations. But the possible combinations are exceedingly limited, for each substance has a specific message, and there is little evidence for insects that different combinations of the substances produce a whole new spectrum of messages. However, an organism could produce a system that does involve the subtle blending of many substances, each nuance having a different meaning.

The difficulty lies in the number of genetic changes needed to evolve such a system, for each new odor would require a separate, genetically controlled system of synthesis of the appropriate substance. With gesture and sound signals, on the other hand, the basic structures involved need not undergo significant genetic change, for a large repertoire of signals can be invented and learned. Once the basic emitting and receiving organs have evolved, production and response to a variety of sound or visual signals, provided the organism is capable of invention and learning, is relatively easy. But even with these capabilities, it is impossible for an individual to invent a new odor; that can only be achieved by genetic change. Therefore visual and auditory signals confer an advantage provided an organism is capable of the kind of behavioral flexibility we associate with innovation and learning, characteristics more evident in vertebrates than in insects.

The speed with which the signal can be emitted and altered is also an important consideration, and again visual and sound signals have an advantage. The reason they are rapid compared to chemical signals is a purely physical phenomenon. It has to do with the speed of light and the speed of sound waves, as compared to the rates of chemical diffusion, even when diffusion might be helped by wind or lesser air currents. Therefore the number of signals emitted per unit time is much greater if they involve showing and calling.

Let us now compare the way in which insects and vertebrates receive signals. In the case of the ant specific chemoreceptors respond to one or a very narrow range of substances. This means that for each chemical signal given off by a nest mate, an ant has a receptor that responds specifically to that signal. There is a one-to-one correspondence between sending and receiving. Consider now the case of Kalmus' police dog tracking people through a field. The dog was able to make extraordinarily subtle discriminations between individuals, and because of the confusion of the identical twins, one is tempted to say that the dog seems to be able to smell our genes. This is a case of recognition; we are presumably not giving off special substances, and the signals are just the natural, individual odors. The astounding thing is the sensitivity of the nose and the ability of the brain to sort out the chemical information received in a meaningful fashion. So one advance in mammals over insects has been in the method of receiving and processing the chemical information. If anything, there are fewer chemical emitters in animals, that is special signal secreting glands; perhaps they are no longer needed because the receiving apparatus is so sophisticated.

The same trend toward more subtle signal reception is also seen in auditory communication. Insects respond to a restricted range of sounds emitted by their fellow creatures; they cannot hear sounds slightly above or below the recording frequency of their ear. An excellent intermediate case in vertebrates is cited in the work of R. R. Capranica (1965) on the mating calls of bull frogs. A male frog will respond only to the call of the same species. By a careful analysis of the sound spectra Capranica has shown that this was a specific response to two sounds, one of low frequency and one of high. If both of these were present the response was maximal. It is also evident from anatomical and electrophysiological studies of the frog's ear, that this is all it can hear. The calls of other species simply do not use these frequencies.[13] However, bull frogs do differ slightly in their call in different regions. These differences are registered by the frog, and they respond more vigorously to their own local dialect. Apparently they hear all the variations in dialect, sort these out in their brain, and show greatest agitation when they hear a call from a frog from their home district. There is both a peripheral processing (what they can and cannot hear) and a central processing (sorting out the variations of what they hear).

The ability of mammals to sort out sounds is truly remarkable.

[13] Furthermore, Capranica (1965) showed that if a bull frog hears sounds over a range of frequencies intermediate between the two frequency peaks to which it responds, there is an inhibition and no response is produced even if the two sensitive frequencies are stimulated.

There may be an enormous racket on the highway; yet a dog will jump up when he hears his master's car. Furthermore the sounds may be quite simple, but because they can be put into a context, they may become rich with meaning. Let me give an example: a car horn has one note; yet it can impart an astonishing array of information to human beings. It can mean: "watch out, I'm coming"; "you just went through a red light"; "move over I want to pass you"; "you forgot to put your lights on"; "it's time to go home"; "thank you for letting me pass you"; "hello fellow citizen of New Jersey with a yellow Volvo"; or "just married." The possibilities seem almost without limit. I can remember playing a salmon on a pool near the road and a number of passing cars stopped. When I finally beached the fish, all the drivers honked their horns, which obviously meant "bravo." We recognize all these immediately, yet the signal itself may be pretty much the same. So it is not only that we hear the sound, but we put the signal in a context that has meaning for us. And clearly we use a variety of other cues as well, largely visual ones.[14]

If we can manage such feats with the single note of a car horn, it is not surprising to me that a chimpanzee can manage with few overt signs or without a complex language. While it is true that the vocalizations of a chimpanzee are hopelessly rudimentary compared to those of man, they are more versatile than an automobile horn.[15] And coupled to the sounds there is a large variety of gestures and facial expressions. In combination the sounds and the visual cues could transmit much information to a brain capable of putting the signals in their proper context and thereby extracting their appropriate meaning.

Mammals therefore not only make use of the most rapid and varied kinds of signals, but they process the information in an elegant

[14] This point has been made in elaborate detail by W. J. Smith (1977) who distinguishes between a *message* that is the signal itself and its *meaning* that is interpreted in various ways depending upon the context.

[15] P. Marler (1976) has an interesting discussion comparing the vocal abilities of chimpanzees and gorillas and considering their vocalizations in terms of the problems posed by their ecology and social structure. The chimpanzee calls fall into thirteen categories, but these can be further analyzed to 343 intermediate calls in a graded series. See also discussion by Marler and others in T. A. Sebeok (1977). Henry Horn has suggested to me that another limitation of chemical signals may be related to the difficulty of providing elaborate contexts to simple signals. Because so many odor signals are not received instantly upon emission, but can be delayed signals, the interpreting of the signal in a complex context becomes a problem.

and complex manner by looking at it in the light of a whole variety of conditions experienced in the past so that they can obtain a profound understanding of the signal. Obviously, not all the information is in the signal; much of it is in the surrounding conditions and much of it is stored in the brain. This storage is the result of learning (the subject of the next chapter), and it is very much part of the basic equipment that has made elaborate culture possible (the subject of the last chapter).

Social animals are communicating animals, and the extent of communication is the one aspect of animal societies that is directly related to culture. Many other aspects are more distantly related to our central theme; but culture is simply not possible without communication.

CHAPTER 6

The Evolution of Learning and Teaching

The subject of learning is larger than any of the subjects we have tackled so far. It is in the province of both the psychologist and the ethologist, and between them they have managed to divide the subject into a number of categories of different kinds of learning. In general the categories represent a range of complexity, and presently I shall give a streamlined version that includes the major steps pertinent to the evolution of culture.

Despite the enormous amount of attention paid to the subject of learning, almost none is allotted to teaching. To an outsider this seems strange because it too has an obvious and interesting evolutionary progression. As we shall see, there is a continuum from the simplest kind of imitation, which may not involve any teaching at all, to the other end of the spectrum where human parents and teachers give complex instruction to a child.[1] When we discussed communication systems in the last chapter, I pointed out that there is a possibility of building in subtleties at both the giving as well as the receiving end. This is equally true of teaching and learning, although there are important differences. In sending and receiving communication signals, we are solely concerned with the emission of a signal and the immediate response to it; by contrast, the process of teaching and learning involves the release of sets of signals specifically designed to alter the behavior of another individual. In the evolution of communication we saw an increasing complexity of the stimulus and the response in higher animals. The same is true in the evolution of teaching and learning. Learning came first during the course of evolution; complex teaching is a more recent invention and obviously basic to cultural evolution.

The foundation of this discussion was laid in Chapter Three where we examined the relation between instinct and learning, that

[1] Many psychologists and ethologists would say that teaching does not occur at all in animals. But this is really a matter of how the word is defined, and the reader is urged to wait for the discussion later in this chapter that shows the details of the continuum between simple imitation and complex instruction.

is the relation between the brain and the genome. Here we shall look at the progressive shift of information transfer by the genome to transfer by the brain in a discussion of the rise of learning and ultimately of teaching behavior as well. However, one obvious point must be made at the beginning. Both learning and instinct, like all living activities, have a genetic component. In the case of instinct, the entire behavioral act is presumably directly inherited, so the offspring can perform the act without any instruction or even trial and error self-instruction. In the case of pure learning, the ability to absorb outside information and to modify the behavior accordingly is also genetically based. Both the inherited automatic behavior and the ability to change one's behavior in response to instructions are presumably inherited in the form of a set of neuronal connections in the brain. The basic brain structure is directly established by the genes, even though, as is true with so many structures of the body, use or disuse can significantly effect further structural changes. Except perhaps in the case of memory, these superimposed nongenic structural changes are minor when compared to the original gene-induced ones.

Learning

There are those who argue that purely instinctive behavior does not exist and that all behavior involves some element of learning. This position was championed by T. C. Schneirla (1953) who pointed out that even an animal in isolation could teach itself essentially by trial and error. He showed that insects, more particularly ants, are good learners. If ants are put in a maze leading to their nest, they soon learn by trial and error how to find their way home. If they are compared to rats in a similar maze, but where food is the reward, the rate of learning of the ants is only slightly inferior to that of the rats (Figure 29). Clearly insects, as well as mammals, can teach themselves.

Learning by trial and error can be considered as an example of negative feedback, as is often pointed out. A correct move is rewarded; a false move is unrewarded (or even invites some form of punishment), and by learning these elementary lessons the organism can quickly manage the correct move. H. S. Jennings (1905) has suggested that certain single-celled Protozoa appear to have something close to learning. For instance, if *Stentor*, which may be attached inside a sheath in one spot, is irritated with a jet of water

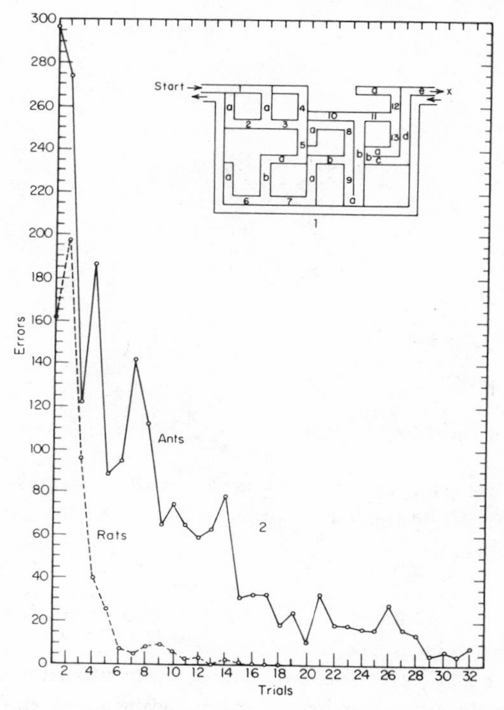

Figure 29. Learning curves for ants and rats. At the top (1) is shown the plan of a maze with six blind alleys. The curves (2) represent the numbers of errors in successive trials for eight hooded rats (broken line) and eight ants, *Formica pallidefulva* (solid line). The rats ran to a place with food; the ants to their nest. (From E. O. Wilson 1971 from T. Schneirla.)

containing noxious carmine particles, it will first try to find a new position, and if this is not successful, it will eventually contract into its sheath (Figure 30). Upon further stimulation it will leave its sheath, swim away, and settle elsewhere. In another set of experiments he disturbed an individual *Stentor* less severely with a gentle stream of water. In this case the first contraction was violent, while the subsequent ones were progressively milder so that ultimately the small, harmless irritation was ignored. There are two aspects to these early experiments: one is the obvious adaptive advantage of

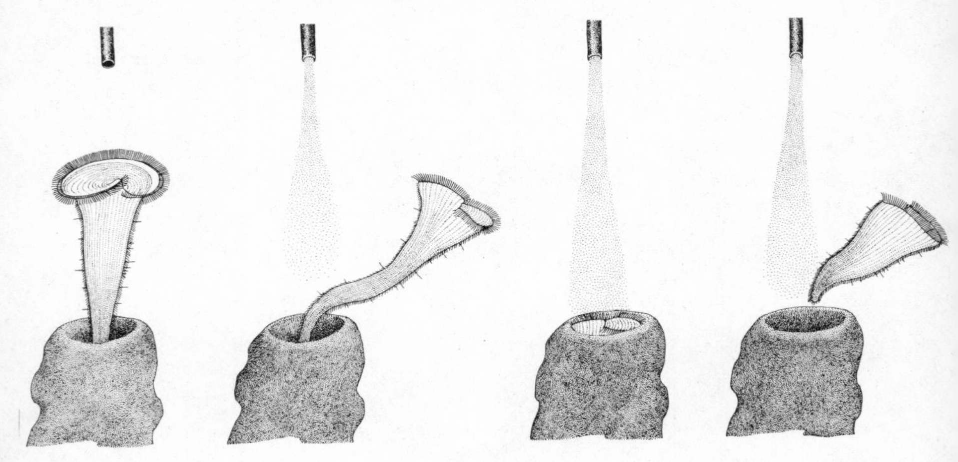

Figure 30. H. S. Jennings' experiment on *Stentor*. A jet of carmine from a pipette is first avoided by the *Stentor*. Upon further disturbance the *Stentor* disappears into its sheath and ultimately swims away.

the organism's behavior, which we will examine shortly, and the other is the mechanism of the response.

There is a danger in this case of entering into an argument concerning the definition of learning and whether or not this should be considered a legitimate example. In the second experiment it is quite conceivable some very simple change is taking place within the *Stentor* that is technically known as *habituation*, that is the lessening of a response upon repeated trials; and if there is simply a fatiguing of the receptors, it is a kind of habituation called *adaptation*. In the first experiment where the *Stentor* eventually moves off, this might be considered a case of *sensitization*, which means that the repeated strong stimulus builds up inside the individual until it finally departs. In both cases the behavior has been modified by repeated, identical stimuli, and one can imagine a continuum between these kinds of simple responses to ones in which there is genuine learning.[2]

[2] It is interesting to compare this example with the situation described earlier on the chemotaxis of bacteria. There if the bacterial cells went up a gradient of an attractant, tumbling was suppressed, while if they went down the gradient, tumbling and the resultant reversal of direction followed quickly. By any stretch of the imagination is there any learning component here? I would think not, because the response is always the same to a particular environmental condition. There is a very specific

The fact that the examples given above are patently simpler than learning in a higher vertebrate might make them useful experimental organisms for the pursuit of an understanding of the basic mechanisms of learning. E. R. Kandel and his group (1975) have used the large neurons of the marine mollusc or sea slug called *Aplysia*, and from their studies they propose a provisional model that helps one to picture the kind of mechanism that could conceivably account for habituation and sensitization. They consider that the crucial activities occur at the end of a sensory nerve that makes a synaptic connection with a motor neuron. This sensory nerve ending stimulates the motor neuron by releasing small amounts of a chemical (a neurotransmitter substance) that crosses the synapse gap to the motor neuron (Figure 31). In habituation (our second case of learning in *Stentor*), the supply of packaged, available transmitter substance is progressively used up, making a weaker and weaker stimulus and therefore a weaker response of the motor neuron. In sensitization a second sensory nerve ending, upon stimulation, secretes the neurotransmitter serotonin (shown in black on Figure 31) into the first. This substance stimulates the primary sensory nerve ending adhering to the motor nerve to produce an excess of its neurotransmitter (acetylcholine) after repeated stimuli, and the result is a large motor response, such as the escape of *Stentor* from the big disturbance in the first example given above. Obviously this model only fits the facts that Kandel and his associates have found from their work on *Aplysia*; a modified model would be necessary for the single-celled *Stentor*. But we are interested in higher forms with a multicellular nervous system, and it may be hoped that work with such invertebrates as molluscs and insects will ultimately produce the needed molecular and mechanistic insight into the process of learning.

The other aspect of our example of *Stentor* is the obvious adaptive advantage of such behavior. The fact that this lowly organism can quickly learn to ignore mild disturbances and avoid in various ways those that are potentially dangerous is a remarkable bit of complex, adaptive behavior in the evolutionary sense. It is of course true that adaptive significance is a key aspect of all learning behavior, high and low, a point that has been emphasized so effectively by the ethologists.

---

and clearly useful response to the environment, but it is totally fixed and never seems to be altered by experience.

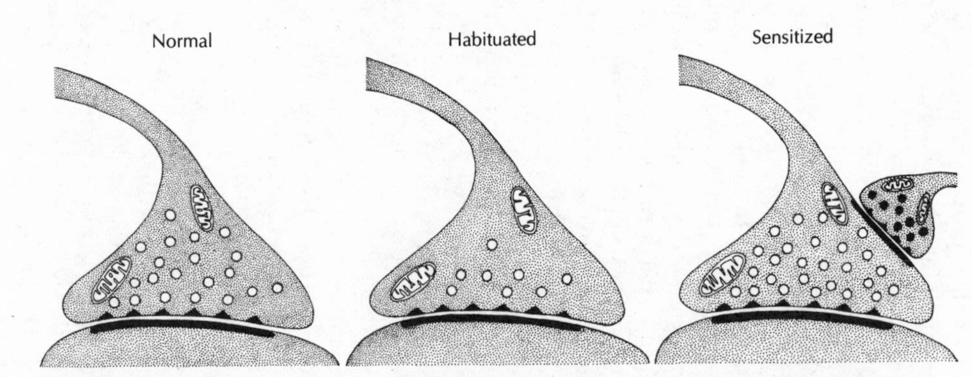

Figure 31. A hypothetical model to explain habituation and sensitization. In a normal synapse vesicles contain signalling substances (white spheres) build up on one side of the gap to the next neuron. In habituated cells, these vesicles become used up and give a lessened response. In sensitized cells the sensory neuron is stimulated by another neuron with a different signalling substance (black spheres) which stimulate an excess of production of the white spheres. (Modified from E. R. Kandel 1976, *Cellular Basis of Behavior*, San Francisco, W. H. Freeman, p. 590.)

No better example of this point can be seen than those from studies on bird song. I showed already in some detail how European cuckoos, cowbirds, and other parasitic birds directly inherit their entire song. The obvious need for such a means of transmission is related to the peculiarities of the life history of these birds. They never see their true parents, and their upbringing is managed entirely by their host or foster parent. Therefore, they never hear the song of their own species; so both the song of the male and the recognition of it and response to it by the female are necessarily innate or instinctive. If any learning were involved, song could not be used as a method of the finding and the uniting of the sexes during the mating period.

Why do not all birds follow this program? Why do some birds, at least partially, learn their song from their parents? Again the answer is thought to be understood in terms of natural selection. P. Marler and M. Tamura (1964) and M. Konishi (1965) showed in the white-crowned sparrow that if the male bird was isolated at hatching, it produced a rudimentary song significantly cruder than that of birds raised by parents. But even this song was not inherited in its entirety because Konishi has shown that if the males were deafened at birth the song they produced was more incomplete and elementary than that of isolated birds. There are, therefore, three levels: the basic song pattern that appears to be directly inherited, improvements by trial and error due to the male birds' hearing themselves sing, and improvements due to hearing mature males sing. It is especially interesting that in many species of birds, including the white-crowned sparrow, there are recognizable dialects of song, and this is one of the things that the males presumably learn from their fathers.

The most obvious adaptive advantage of learning song is that it enables birds to recognize individuals by small differences in individual song or groups of birds by differences in dialect. Learning permits the song to be more variable, more flexible, and this is essential in the identification of other members of the family, and, in the case of dialects, the local population. An extreme example of this hypothesis is a suggestion put forth by J. L. Brown (1975) to explain the extraordinary ability of parrots, mynahs, corvids, mockingbirds, and starlings to imitate sounds, including human speech. He argues that perhaps this ability is associated with a strong pair bond because all the species mentioned above characteristically have close links between mates. The assumption is that they learn to imitate one another so that they have private sounds between them that help in locating one another and play a role in other acts of togetherness.

A variation of this theme can be found in the remarkable phenomenon of duetting found in some species of birds, such as various African shrikes (Hooker and Hooker 1969; Thorpe 1972). The male and female may sing the song together in perfect unison or the alternate notes are given successively by each member of the pair with such rapidity and precision that the result sounds like the song of a single bird (Figure 32). Duetting birds are characteristically found in thick forests, and it is presumed that the double song is a way in which they can keep in touch with one another in dense foliage. These duets involve considerable learning during the initial practicing period of the development of the song. It is not so much a case of individual differences, as it is learning a harmonious song arrangement with a mate that not only strengthens their bond, but also permits this bond to hold when they cannot see one another.

Let me now give another example of behavior in birds where there is a mixture of learning and direct inheritance of a pattern. This one is particularly interesting because the learned component also directly leads to a cultural transmission. The migration pathways of some birds are known to follow certain geographic routes, and there is some evidence that this is learned by younger birds following and imitating older birds. In S. T. Emlen's (1975) work on indigo buntings he has shown that since this bird migrates by night, it can make use of the stars for its navigation. He did his experiments using circular cages placed in a planetarium and was able to establish the direction the birds hopped by placing them on an ink pad and recording their foot prints on a funnel shaped circle of blotting paper (Figure 33). He obtained evidence that the buntings had no time sense, but recognized patterns of stars in the Northern sky. The ability to recognize these star configurations is acquired during the nesting period; young birds raised in diffuse light were unable to orient in the night sky when they became adults. Apparently they learned the sky by watching the stars rotate around the North Star during the course of the night.

As adults they can tell the North from the South in a fixed sky; the rotation is important solely for the learning process but is not needed for North-South direction determination once the configuration of the stars is imprinted. Even more remarkable is the fact that the question of whether the bird flies North or South depends

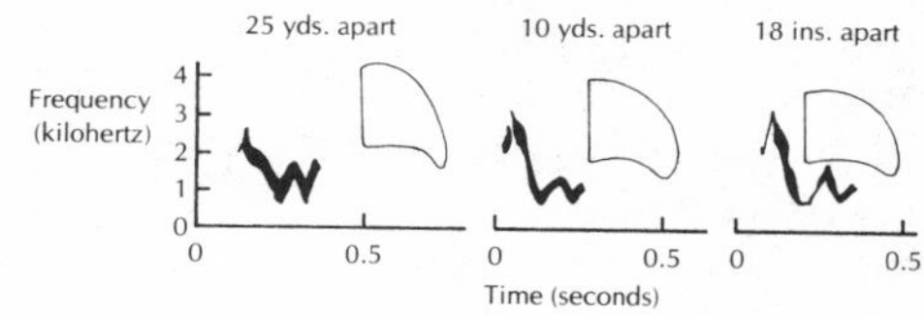

Figure 32. A pair of duetting African shrikes (*Laniarius erythogaster*). Because they inhabit dense foliage they sing a duet, as can be seen from the sonogram of both birds below. Note that the timing of their antiphonal song varies depending how far apart they are from one another. (The sonogram is after T. Hooker and B. I. Hooker 1969.)

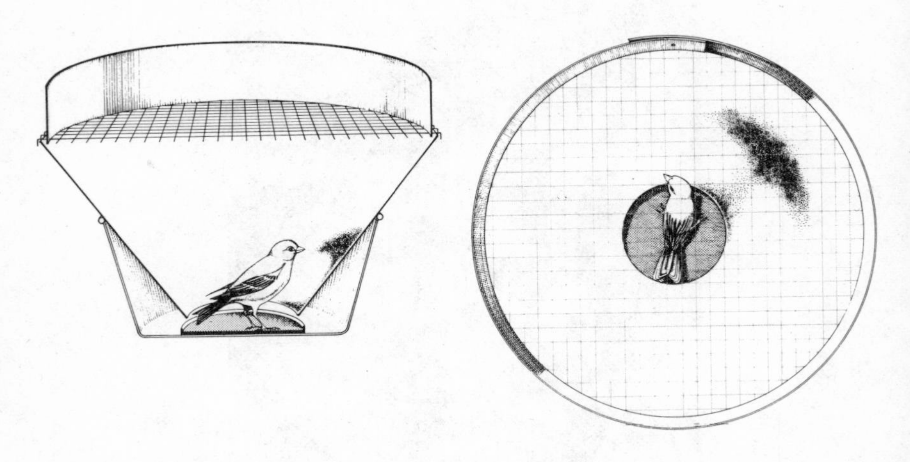

Figure 33. Circular test cage for determining the directional preference of an indigo bunting is shown in cross section (left) and top view (right). White blotting paper is on the funnel part of the cage. The bunting stands on an ink pad and each time it hops onto the sloping funnel it leaves black footprints. (From S. T. Emlen, 1975 courtesy of the *Scientific American*.)

upon its hormones. In a spring day and night cycle with lengthening days the reproductive system becomes active and the bird flies North. But if a bird is tricked by giving it autumn day and night cycles with a shortening of the days, then it will fly South. The polarity of its innate response is determined (or reversed) by hormonal changes, while its image of and understanding of the sky is learned by spending its early youth staring at the rotating night sky. After a bird adds what it has learned about flight routes from older birds, its behavior pattern is an intricate mixture of innate and different kinds of learned responses.

In this example of bird migration and the previous example of bird song, one feature of the learning should be stressed. In both cases some of the learning takes place during an extended youthful period when this kind of learning is especially or solely efficacious, and this immediately suggests that the process is a kind of imprinting. In classical imprinting there is a relatively restricted period when it is possible for the young animal to learn; a stimulus received afterwards is ignored. Because imprinting involves a critical period during the growth of the animal, the learning is necessarily slow, as are all developmental processes.

However, other kinds of learning are rapid and may even occur after one lesson in adult individuals. In both the slow and the rapid learning there could be adaptive explanations. In the case of quick learning the adaptive advantage is often easy to see. There are numerous examples where one lesson teaches an animal to avoid a certain food; for example, a bird avoiding a distasteful insect. The experience may be unpleasant and possibly even dangerous, and it is optimally efficient for the organism to find this out as quickly as possible.[3] Another example might be confrontation with a dangerous predator. A narrow escape will teach the prey to avoid henceforth all similar situations.

In the case of the slower learning, some instances may simply be slow because there is no hurry; but more likely the need for a prolonged period of learning is due to the fact that what is learned is difficult to absorb and yet vitally needed. This might well be true of learning the night sky for indigo buntings, or it could be so for any action that needs to be done with exceptional precision. Imagine, for instance, the muscular dexterity needed for sudden escape from a hawk by a monkey. This has been described by S. L. Washburn and D. A. Hamburg (1965) who point out that the extended period of play fighting may be essential for their ultimate survival. It is only after an enormous amount of practice that they will be able to make the formidable escape leaps in a fraction of a second, when a hawk swoops down among the trees. Since any error will result in death, the selection pressure for a long and careful apprenticeship will obviously be very strong. Escape for these monkeys involves both kinds of learning: the slow practice is needed for the extraordinary muscular dexterity and the initial escape response is likely to have been learned very rapidly.

Memory

The efficacy of learning depends to a large extent upon the capacity of the organism to remember. Therefore, besides the question

[3] This quick learning has been shown for numerous animals including both invertebrates and vertebrates. It is known as the Garcia effect and it manifests itself by associating a particular food taste with some unpleasant body discomfort. In vertebrates nausea is a common form of such discomfort. For an excellent discussion of the phenomenon see P. Rozin (1976).

of how much and how complex the information can be that is learned, there is the question of how long it is retained. In fact because there are so many elements that go into the learning process difficulties arise in giving a simple and straightforward description of it.

The mechanism of memory is poorly understood. All the early work was done with vertebrates, but more recently there has been a flurry of activity with invertebrates. Of special interest is the work on insects, some of which will be mentioned further on. In all the animals examined so far a rough separation seems to exist between short term and long term memory. Perhaps the most exciting new insight is that insects and vertebrates alike have a special region of the brain, separate from the storage area itself that is involved in the consolidation of the learned information. The evidence is clear-cut. If this region is anesthetized or removed, there is no memory storage. If, however, the region is put out of commission after storage, then the memory remains even without the consolidation center. What is not known is how this consolidation into storage is achieved, that is, its physical and chemical mechanism.

One way to investigate this important problem is by the use of a relatively simple organism, such as an insect, to dissect more readily the components of learning. For instance, as previously mentioned, R. Menzel et al. (1975) have trained bees to respond to olfactory cues while they are fixed in one position with their brains exposed. If the lobe at the base of the antenna is briefly inactivated by cooling with a fine wire, the learning is totally blocked, but only if the cooling is done within one and one-half minutes after the lesson. If the lobe is chilled at two minutes or more, the memory is unimpaired. Other specific parts of the brain can be cooled at later periods to achieve the same result. Clearly the information moves to different parts of the brain at different times. W. G. Quinn and Y. Dudai (1976) are pursuing another approach using mutants of the fruit fly *Drosophila* that are deficient either in their ability to learn or in their ability to remember what they have learned. (Different mutants differ in the degree of learning and in the timing of the memory losses.) By the use of the mosaic mapping method devised by Sturtevant (1929) and refined by Benzer (1973) and others more recently, it should be possible to find what part of the nervous system has the malfunction. Hopefully these approaches, coupled with more direct neurophysiological and anatomical ones, will lead to a deeper understanding of the mechanisms of learning and memory.

Teaching

As I pointed out in the beginning of this chapter, I can nowhere find a systematic study of the subject of teaching, which is surprising when one thinks of the excessive amount on the subject of learning. Here I shall give no more than a brief outline comparable to what has just been said about learning. This outline will present the different complexities of teaching and the progress of these complexities over the course of evolution.[4]

Perhaps the most primitive kind of teaching is self-teaching. It is the same thing as what we have just described as trial and error learning in the case of birds that can hear their own songs or monkeys that learn how to make great leaps through the trees by long practice. In these examples no information is transferred from one individual to another, which is the kind of teaching that is necessary for the passing of cultural information. This does not mean that self-teaching is solely a property of primitive organisms; all mammals including ourselves indulge in a large amount of self-teaching.

The simplest form of one animal teaching another is hardly teaching at all. It is teaching by example when one individual imitates an older and more experienced friend or relation. The teacher does not actively impart special information, but simply carries on its tasks in the presence of other less knowledgeable individuals. This kind of teaching is probably the most commonly found in the animal kingdom. It reaches great heights in the higher vertebrates, and we already have seen some instances. The imitation of the song of a bird by another bird or the majority of the learned actions of young primates during their long period of parental care come under this category. Even among human beings, not only do small children imitate adults, but the very word apprentice to some extent implies imitation and practice, as opposed to the word student that implies receiving a set of instructions or a quantity of information.

Since vertebrates are so good at imitation, it is not unexpected to find examples where the young are occasionally reprimanded for doing a poor imitation. The parent will supervise the fidelity of the imitation, and if it does not come up to snuff, it will nudge, poke,

[4] I have purposely refrained from giving numerous examples of teaching among animals to save them for the last chapter. Each example of teaching is also an instance of transmission of cultural information.

or slap the offspring to encourage a better performance. Many examples of maternal attentions of this sort have been observed among primates as W. A. Mason (1965) points out. Such behavior is exaggerated in captivity perhaps because the mother is freed from the normal pressures of food gathering and protection and gains stimulus in a dull environment from her offspring. He quotes a study from the Yerkes laboratory in which the "infant is inhibited, curbed, directed, driven, or encouraged in multitudinous ways by maternal attentions" (1965: 527). Although often muted, the same behavior is found in the wild. Note that the parental guidance is both positive and negative. The amount of information that the parent passes on in this manner is very small, but by behaving in this way she makes use of the powerful ability of the young to learn by trial and error. In effect the parent is reinforcing the self-teaching, and this is a quick and effective way of guiding behavior.

To me the surprising thing is that such a large share of teaching by higher vertebrates falls into this category. The ability and the tendency of the young to imitate is high, and the ability of parents and other adults to guide the imitation is easy for it involves very little communication. It can be done by the simplest kinds of gestures or physical restraint, and if a sound is used, it may be a simple one that means yes or no. J. Goodall (1965) gives an example used when infant chimpanzees have climbed too high in a tree. The mother taps the trunk softly and the child climbs down at once. It is hard to imagine that this kind of teaching will be very important in the passing of cultural information; yet as we shall see, it does occur.

This leaves the last and most interesting category of teaching in which one individual passes an appreciable amount of information to another. As was pointed out earlier, there is a continuum from no information passed through the simple reinforcement we have just discussed to the extreme case of complex teaching and instruction in man. If we look to the most phylogenetically primitive cases of this third kind of teaching, we find them in insects, and more specifically in the great work of K. von Frisch (1967) and others on the language of the honey bee.

The very fact that it is referred to as a language directly implies that teaching is involved. In this case a foraging bee enters the hive and gives off a series of signals that turn out to be very specific instructions to the other worker bees. Since it is now an old and well-known story I shall only summarize the extent of the teaching.

In the first place the scout bee tells the other bees she has found a source of nectar by her frenzied activity when she returns to the hive. They may also smell the flower scent on her, and she will regurgitate some of the nectar; however, this handing out of samples is not teaching, but supplementary to it. Secondly, she tells her fellow worker bees the distance to the food source. For short distances she does the round dance; for distances over a certain threshold, depending upon the variety of bee, she does the waggle dance (Figure 34). The rate at which she does the straight run of the dance bears a linear inverse relation to the distance; in other words the slower the dance, the greater the distance. From the numerous experiments of von Frisch and others it has been established that this message is very accurately interpreted by the worker bees who fly out on their own without anyone to imitate. Finally the scouts instruct the other bees as to direction only if the nectar is far enough away for the waggle dance. The cue is the straight part of the dance, but the extraordinary thing is that both the scout and her pupils can translate a gravity direction into a visual direction. The dance in the hive is normally in the dark on the vertical surface of a comb. The code is very straightforward: a dance straight up means toward the sun, and a dance straight down means away from the sun. But if the dance is 20° to the right of the vertical, this means to both the teacher and the student that the nectar lies in a direction 20° to the right of the sun (Figure 34). In this way all the points of the compass can be transposed from gravity to sun position. It is a remarkable feat quite beyond our own powers: we could never sense gravity to this fine degree, nor would we be very good, without a few mechanical aids, at walking at a fixed angle to the sun. But honey bees can do just this and therefore transmit accurate and useful instructions from one individual to others. This is an instance of genuine teaching.

Von Frisch's discovery of these capabilities of bees tempt one to consider bees quite as exceptional as man in their own special way. However, this might be seriously wrong, and it is quite possible that we have yet to learn about the specialized languages of many organisms, both invertebrate and vertebrate. Each case is rather like cracking a code, and few people have this gift; Champollion, Ventris, and von Frisch are conspicuous exceptions. Let us therefore assume that similar kinds of teaching might be reasonably common, and consider bee language more closely as an example of a method of complex instruction.

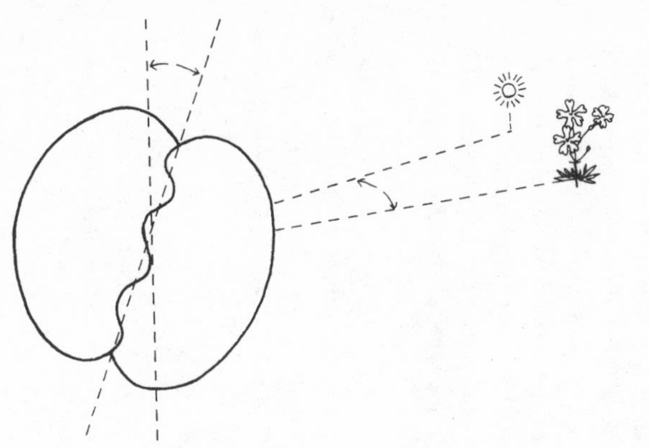

Figure 34. The waggle dance of the honey bee. As the bee passes through the straight run she waggles her abdomen, and then returns alternately on one side or the other in a circle to the beginning of the straight run and repeats her dance. If the straight run is 20° to the right of the vertical in the dark hive this tells the other bees that the source of nectar is 20° to the right of the sun as they fly out of the hive (see above).

The most obvious thing for a human being to say about bees is that they are relatively limited in the number of instructions they can transmit. In such a statement is clear that our own capabilities have been the yardstick and, by the same token, we cannot help but admire the precision of the statements of the bee instructor. It is not just that there are so few instructions, but that they are so limited, so specific, and so precise. The rigidity of the teaching is quite unlike anything in our own experience. On the basis of this one example one might conclude that teaching in insects arises in very special, narrow circumstances, and the amount of information transmitted is confined to a few directions for one act of behavior. This fits in with what we know of insect learning behavior. Ants, for instance, can learn mazes with considerable skill, and it has been proved that insects are capable of memory. These skills are roughly what is required if an insect is to receive instruction of the sort handed out by scout bees. Perhaps what insects cannot do to any appreciable extent (and this is the quality that possibly separates their learning behavior from vertebrates) is to imitate.

It is just this lack of the power of obvious imitation, more than anything else that delimits the character of insect teaching. It is only possible to give instructions when both the instructor and his pupils have an inborn, innate ability to give and receive. What bees inherit is the code for distance and direction, the ability to send the code, and the ability to read it.[5] The whole procedure provides the social group with an effective means of gathering nectar, giving them a decidedly selective advantage over any less well-organized nectar gatherer. But what flexibility exists lies entirely within the confines of questions of distance and direction. Indeed many directions and distances are possible, but there is no teaching beyond those two types of information. For instance, a scout bee has no way of telling the other bees anything about the color of the flowers or their shape and size. But most significantly of all, if anything unexpected happens, she has no way of communicating it. As an example, if the flowers are on a horse-drawn cart that is slowly moving west a scout certainly will not be able to transmit this fact, nor is it likely that she understands the peculiarities of the situation herself. The rules of the behavior pattern are very strict, and there is no chance for any

[5] Yet, at the same time, there may be a considerable amount of learning in the details of acquiring the code itself (for example, Lindauer 1960).

deviation from those rules and no chance of any kind of innovation unless it be a slow genetic one.

If we turn now to higher vertebrates, especially birds and mammals, the contrast to insects, even clever ones like bees, is sharp. First of all, to catch a thread from the last paragraph, vertebrates have remarkable powers of imitation. It is not only that they can readily imitate, but they are highly motivated to do so. Active imitation is the stepping stone to the more elaborate teaching in the vertebrates; without it teaching never would have developed beyond the level found in insects. Because of its importance, let us examine imitation again, but now we want specifically to examine its adaptive significance, along with that of teaching.

Imitation is a process we associate particularly with the period of youth, and the key to assessing the degree of imitation involved in the development of any one organism is seen from experiments in which the young are isolated from birth or from hatching. T. C. Schneirla (1971) gave an example with army ant workers: when they emerge from the cocoon in isolation, they can perform the functions of the other workers, but must take an extra day to be able to manage all the skills. This must involve either a negligible amount of imitation, or it could easily be explained as the basis of none at all and that the presence of other workers simply stimulates the more rapid development in nonisolated workers. The biggest advance is found among many lower vertebrates and all higher ones in their period of parental care. During this period, not only are the young constantly at the side of one of their parents, but the parent is actively doing things important for survival; it is, for instance, eating or escaping from predators. My first premise, therefore, is that imitation became a significant process during this period.

In parental care, it is advantageous for the parent, in its effort to contribute genes to future generations, to protect and watch over the offspring as solicitously as possible. This is the first evolutionary step toward teaching, based on obvious and well-established principles. Equally important for the same reasons, it is advantageous for the offspring to protect themselves, and if they can do better than just to rely on the parents, they will correspondingly increase their chances of survival. Since parents have mastered all the special survival skills, it is obvious that the sooner the young can imitate their behavior, the safer they will be; therefore there could conceivably be a selection pressure to imitate. It is true that many protective responses to predators are directly inherited. Chicks of many birds, especially ones that scratch round in the undergrowth following an adult like partridges or sandpipers, will respond to the appropriate call of their mother by either freezing or scurrying to cover. But there are many other aspects of either escaping predators or obtaining food that are learned, and a large part, if not all of this learning is by imitation.

If the imitation can be improved by the parent's reinforcing the trial and error process, this is also clearly advantageous. If this simple approval or disapproval instruction on the part of the parent can turn into signals conveying more information, again it would be encouraged by natural selection. For instance, if a bird or a monkey could produce different calls for different emergencies and if these were not inherited but learned responses, then the animals would be specially protected for two or more different kinds of emergencies. The fact that some behavior is learned rather than innate in the young could also be adaptive. For instance, one species might encounter different dangers in different geographical regions, and a more flexible, learned response would allow it to modify its behavior in a manner appropriate to the region where it is reared. I do not want to anticipate the discussion of the next chapter, but evidently if the behavior is not fixed in the genes and if it concerns the well-being of the young, the teaching of the parent and the learning of the young will both contribute to the likelihood of the offspring's survival.

According to the views I have expressed here, the first real teaching, over and above the simple reinforcement of the offspring's ability to imitate, would be found in instances where a parent will give its young instructions about protection or food. From this moment onward, the extent of the teaching is totally limited by the ability to communicate, the ability to make use of a language. This may first have been a few sound signals or gestures or a combination of the two. But these signals are not necessarily, in themselves, repeated or imitated; they are acted upon. It is true, as we have just said, in primitive examples, such as some bird alarm notes, the same thing could be achieved by instinctive calling and instinctive responding. But in some species it became helpful to have the response to the signal learned, taking advantage of the vertebrate's innate ability to imitate; and once a signal needs to be learned, giving that signal becomes a more advanced form of teaching.

We do not yet know what the language limitations are for other animals besides ourselves. There may be every conceivable shade of intermediate kind of communication among animals, but we do not understand their language and therefore miss it entirely. That such unknown languages might exist is suggested from the work on chimpanzees with sign or computer language that has found these apes can communicate a large amount of information if we teach them some kind of manual signalling. This raises the question of the kind of communication involved in teaching. We tend to think that auditory signals are the principal means, because they are for us. Clearly they may also be for other animals, such as birds and more especially dolphins and whales. But there are numerous other possibilities. Visual information and even tactile signals could be used in animal language and could be the means of teaching. More important, any combination of all three of these could comprise the total methods used. We certainly use them all, even though our auditory signals play by far the greatest role. But even deaf persons can converse rapidly by sign language, and deaf and blind people can be taught, as was Helen Keller, by tactile signals. One could imagine that since chimpanzees learn so much when taught with sign language, that they somehow do communicate and teach one another by a combination of signals that we cannot yet fully interpret. If this is so, we have not cracked their code any more than we have for the majority of animals, high and low.

Besides teaching by using the direct signals, we can also teach indirectly by creating artifacts. For man the most obvious is writing, although in this modern age there are others that involve electronic devices, tape recorders, television, and so forth. In writing we put the lesson down in a form that corresponds to speech, and in prose we can describe gestures and facial expressions so that words can convey a great depth of meaning in the lesson to be learned.

In less sophisticated ways animals also leave artifact messages. A simple example would be instructing by leaving a trail; ants leave trail substances that they secrete from one of their many glands. But here we are no longer talking about significant teaching. The ant that follows discovers only that another ant has been there, and she instinctively runs along the trail; this is a signal rather than a lesson. As is so often the case, there is a continuum between signals and lessons, but in all instances where artifacts are constructed by nonhuman animals there is very little if any teaching. In those cases where nests are constructed and elements in the skill of construction are learned, as in the sleeping platforms of apes; the young will not learn how to make a nest by admiring the final product, but rather by watching and imitating the parent building the nest.

Because of the use of writing and the extraordinary richness of our spoken language in teaching, man is clearly in a very special class when compared to all other animals. The evolution of learning was gradual, and with all animals the ability corresponds roughly with the relative increase in the size of the brain. Extensive, complex teaching, on the other hand, not only appeared much later, but the degree and the sophistication of teaching appears to have increased very sharply somewhere in the early history of man. We cannot yet be too specific as to when and what the shape of the rising curve might be because we still know so little about teaching in primates and other vertebrates as well. It, along with the meaning of animal signals in general, will be one of the most interesting areas of future research.

Summary

It may be helpful to summarize briefly the progression of the levels of teaching and their relation to learning. There have been, among different animals, changes in both the sending and the receiving end of the process during the course of the evolution of teaching.

At the receiving end the most primitive condition, found especially in insects and other invertebrates, is a fixed response of an organism. This is sometimes called a fixed action pattern and examples are legion: orientation to light (phototaxis), contraction or escape upon being touched, the shooting out of the proboscus of a fly in the presence of sugar water, and so forth. A more advanced condition occurs when the organism can respond in more than one way, but again each alternative is rigid. This might be a fly moving toward or away from light, depending upon its intensity, or the fight or flight responses of a mammal. The third level includes those cases of a slight flexibility in the response by being continuous over a range of related but quantitatively varied stimulus. A good example would be the honey bee's response to signals indicating different degrees of distance or direction. In the fourth and most important level an individual has to *learn* the correct response to a stimulus rather than give an automatic, innate, fixed response. The line between level three and four is not sharp, for M. Lindauer (1960) has shown that the honey bee learns the direction in which the sun moves, so there is some component of learning in her flexible response to direction. Vertebrates characteristically show a heavy component of learning in their response to signals. Learning is involved in those cases where the response is by imitation, and in more advanced situations where there is a response to complex instructions. Furthermore there is a wide spectrum in the kinds of learned response from a simple pattern of behavior as in the honey bee to a complex skill, such as the technique for effective prey catching or foraging in vertebrates.

In the evolution of teaching, the changes in the nature of the signal containing information are particularly relevant. Again the most primitive case is a single signal given off by one individual and received by another. This is found in all groups of animals from primitive invertebrates to advanced mammals. Next there may be more than one signal, each one of which gives different information for different responses. As one moves up the evolutionary scale, there is a progression toward a great proliferation of these signals that become increasingly loaded with information. The signals can multiply by simultaneously using more than one means of sending (sound, visual displays, odor), and these can be combined in various ways. Or they can proliferate, as they did in early man, by an increase in the signals from one sending method (sound) and produce an elaborate language. Finally, it is possible to put the language into some form of artifact such as writing, and in this form it can effectively transfer information for the purpose of teaching.

CHAPTER 7

The Evolution of Flexible Responses

The discussion so far has been concentrating heavily on two aspects of behavior: social interaction that evolved with more effective means of communication examined in Chapter Five and the process of learning and teaching in which evolution permitted a progressive increase in the amount of meaningful information that could be transmitted from one individual to another examined in Chapter Six. In this chapter I shall return to the basic genome-brain dichotomy and concentrate on the flexibility of responses and their evolution, which involves both the genome and the brain.

Culture and its transmission is the ultimate in flexible behavior, while heredity and gene transmission is at the opposite pole since it is relatively fixed and rigid. Animals with cultural evolution learn new information and pass it on by teaching and are capable of innovations and inventions; this exists in our own species. In genetical evolution the Mendelian rules of transmission are set and even an innovation in the form of a mutation or a chromosome rearrangement must obey those staid rules. But these two conditions are the extreme opposites and there is much in between. The genetic system can be more flexible and the behavioral system can be more rigid. We shall examine this gamut from rigidity to flexibility in terms of its evolution. The presumption is that since culture needs flexibility in order to survive and evolve, the trend toward its increase plays a key role in the development of the cultures of various species including man.

Genetic Flexibility

The genetic system has changed to increase its flexibility in a number of ways. Since this subject is somewhat remote from the other end of the spectrum, namely culture, it will be examined briefly and not in any exhaustive fashion. The manner in which variation is controlled through mutation and recombination is in itself a step toward flexibility. The continuum starts at the rigid end with asexual, vegetative, or parthenogenetic reproduction in which there is no variation; asexual reproduction with mutation such as one finds in bacteria and many other microorganisms; and finally sexual organisms where there is the advantage of both mutation and extensive recombination. The resulting variation is a means of inventing new survival machines that can manage either unexploited environments or that can cope successfully in a changing environment. The flexibility provides increased options for stable and successful existence in a world that for physical and biological (and cultural) reasons never seems to stay still, but is always shifting, often in radical ways. Presumably this explains why all organisms do not reproduce solely by vegetative budding or parthenogenesis. Asexuality is a successful strategy in a constant environment, but may be wholly lethal in a changing one. By contrast the sexual system, with its great store of genes and its ability to put these out in new combinations, can create survival machines capable of coping with new and unexpected exigencies of the environment. One assumes that one reason sexuality is ubiquitous among all animals and plants is its success in supplying the flexibility needed to meet the external changes. It has probably been selected for simply because of the environmental fluctuations.[1]

The environment can change in time and also show complex local differences in any one area; it is a time and space patchwork. Besides directing the appearance of straightforward variants, the genome has produced a powerful way of being flexible in polymorphism. By this we mean that an animal or a plant can exist in more than one genetically determined form or *morph*. The simplest type is sexual dimorphism or a developmental polymorphism.

In those species that show sexual dimorphism it is possible to understand differences between the sexes in terms of Darwin's concept of sexual selection: one sex competes for the attention of the other and this competition results in individuals of the competing sex becoming different in some way, usually in size or in color patterns. However, there are cases where, over and above the sexual selection, there is an avoidance of competition between the two sexes. That is, the sexes will become so different that they eat different foods and thereby exploit the resources of the overall envi-

[1] The matter of the evolution of sex is something of great interest, and there are numerous discussions. The writings of G. C. Williams (1966, 1975) are especially useful. See also the recent important book of J. Maynard Smith (1978).

ronment more effectively. There are a number of examples of this; for instance the bill shape and feeding habits of some birds are as different between the sexes of one species as one might expect from individuals of two separate species (Figure 35. C. Darwin 1874; D. Amadon 1959).

Many species of animals have radically different morphs at different stages of development, and the young individuals do not compete with the adults for food. The most common examples are found among insects where the larvae may be aquatic or terrestrial and the adults winged; the young and the old of the same species inhabit two totally different worlds. Note that this example involves time as well as different feeding niches.

Over and above these two examples of rather elementary and obvious polymorphisms genetic polymorphisms sometimes involve many more than two morphs. A classic example is found among various species of butterflies in which the different morphs mimic butterflies with an unpleasant taste to bird predators (Figure 36). An extreme case is an African species of butterfly that mimics

Figure 35. Extreme sexual dimorphism in the New Zealand Huia (now extinct). The male (above) and the female (below) feed on different kinds of tree-dwelling insects. (By Margaret La Farge. Reprinted by permission of Houghton Mifflin Company from *Watching Birds* by Roger F. Pasquier. Copyright © 1977 by Roger F. Pasquier.)

Figure 36. Mimicry in African butterflies. The models on the top are three distasteful species of butterfly. The mimics below them are different female morphs of one species, *Papilio dardanus*. They are edible, but gain protection by looking like a distasteful species. (After a photograph of V. C. Wynne-Edwards 1962, *Animal Dispersion*, Edinburgh, Oliver and Boyd, p. 442.)

thirty-three separate distasteful species, so that it has thirty-three morphs (G.D.H. Carpenter 1949). Snails often have polymorphic color patterns, and there is evidence that at least some of these patterns are adaptive and that the same species may be able to have a high density in an area of mixed vegetation by having the different morphs invade different kinds of habitat. In this case it is not that a snail or a particular pattern necessarily seeks one habitat (although this is theoretically possible); rather it is that the predators cull different areas in different ways, depending upon the effectiveness of the cryptic coloration of the different morphs. Culling can also occur in the same area at different times. For instance, the fall and the spring coloring of the vegetation will favor different cryptic coloration in the animal; the morph that is best hidden from predators during any one season will obviously be the most abundant at that time.

Let me give one more example of a complex genetic polymorphism. In a moth in Switzerland that damages cabbage crops two morphs alternate in successive population explosions (L. R. Clark et al. 1967). When one of these reaches its peak, it is attacked by a parasitic wasp infecting almost the entire moth population. The resulting population crash is not total, however, because there are some individuals of the other morph that are relatively resistant to the parasite. This new morph now builds up a large population rapidly, but it is susceptible to a virus infection that infests all the individuals except a few of the first morph that are somewhat virus resistant. So the two morphs of the moth cycle rise and crash more or less alternately, although there are always some individuals of both in the population. In this case there is a clear temporal effect involving the bloom of two separate parasites.

As J. Maynard Smith (1975) points out, one can account for these instances of balanced or stable polymorphism in three principal ways, and all of these explanations involve selection. In the first, which would apply to the snails discussed above, selection favors certain morphs at different locations or at different times, and if the habitats and the seasons appear regularly over a long period, then the polymorphism will remain stable. In the second case the most fit morph is the heterozygote with an unlike pair of genes at one locus (*Aa*). If this heterozygous morph is selected for more intensely than either homozygote (*aa* or *AA*) then it is assured that both genes (*A* and *a*) will remain in the population and the polymorphism will be stable. The third way of maintaining a polymorphic state is to have a gene become more fit as it becomes rarer. An example of this again comes from the feeding habits of birds and fish. While feeding they tend to keep eating the same color pattern, as if they kept in mind a model of the ideal food, and by doing this they concentrate on the more common morph and ignore the rarer one. In these ways the genes for all the morphs may stay locked into a population for long periods of time. Nevertheless, as we shall see, the flexibility of these polymorphic systems is severely limited compared to nongenetic flexibility; it can only cope with a small, finite number of environmental changes.

Some organisms accomplish the same kind of restricted flexibility in a totally different way not involving gene differences in the genome, but rather the manner in which the gene expresses itself. One of the oldest accepted notions is that genes and gene products are influenced by different environments. In modern terms this means that the transcription of the gene to the corresponding messenger RNA or the translation of that message into protein in the cell is controlled by factors in the cytoplasm. It is also possible that the proteins synthesized are affected by cytoplasmic factors. In other words, the gene activities within the cell wholly depend upon signals from the cytoplasm, which may in turn be affected by conditions of the larger, outside environment. This being the case a system may develop in which an organism (or parts of an organism) can take on different morphs depending upon the environment. I have already given two examples in Chapter Four: in one case the leaves of an aquatic plant, such as the arrowhead, have a totally different morphology depending upon whether they grow in air or are submerged under water (Figure 13); in the other case the pigmentation of the Himalayan rabbit depends upon the environmental temperature.

There are other examples of this kind of effect (which has been called phenotypic polymorphism). For instance, in the swallowtail butterfly in Sweden the pupae have two color phases, green and brown. In the Southern part of the country there are often two generations a year, one maturing in the summer and one overwintering in the pupal state. The former have predominantly green larvae, the latter brown. C. Wicklund (1972, 1975) has shown that these color differences are primarily due to the wavelength of light reaching the larva during a sensitive period before pupation and that, at least in the summer generation, the green color did confer a selective advantage over the brown pupae because they are harder to see in the green summer vegetation.

Primitive Behavioral Flexibility

In the cases discussed so far there is little or no behavioral component; the flexibility results entirely from an interaction between the environment and the genome or its immediate products. We shall now examine a variety of examples of flexibility that seem to involve both the genome and the brain. As is characteristic of all such instances, it is impossible to determine, or even measure, the relative genetic and behavioral components; it is only clear that both exist. One of the reasons for the uncertainty is that the two are so difficult to separate. All behavior, no matter how flexible, has a genetic basis, making it hard to determine when an action becomes purely genetically determined. In fact, this is a rather meaningless question to ask, for as I have just pointed out, we only know genes as a result of their actions in the survival machine that they influence or control. If, for instance, a set of gene products is responsible for specifying a particular reflex action to a specific external stimulus then we can begin our unrewarding questioning of whether this is to be considered genetically or behaviorally determined. Obviously it is both. Now consider the next step when not just one behavioral response is possible, but two; which of the two the animal makes is determined in the brain. To make the decision the animal may have to take a number of factors into account, and the brain can do this and decide upon the most favorable response. But the computer and every bit of the neuronal machinery wired together in the brain has presumably first been mapped out in the genome. So the decision-making process has as firm a genetic basis as the simple reflex arc. However, there is an important difference. The neuronal machinery involved is more complex in the second case; specifically this elaborate mechanism is used to make a choice between two alternatives. The flexibility of the response therefore comes as a result of gene changes that produce the complex brain.

The important distinction between a reflex action and a brain-mediated decision is that the former has one response only, while the latter has two or more. It is true that some of the examples I gave in the previous section also had two responses, such as the environmental determination of the leaf form of arrowhead, the water plant. But this kind of example is at another point in the flexibility continuum. So to make my meaning clear, in order to consider a response behavioral the choice between two or more responses must be made by the brain or in other parts of the nervous system. Even the sense organs may give off alternative impulses to the brain depending on their response to different external stimuli. Therefore, instead of referring to an act as being genetically, as contrasted to behaviorally, determined, it would be better to say that the nervous system response involves either single or multiple choice. It may be difficult for all of us to wean ourselves from the older terminology, but it would be far more meaningful to say that an instinctive, innate, fixed response is simply a *single response* behavior, and all those cases we call behavioral are *multiple choice* behavior.[2] In the latter case true learning is possible, while in the former case one could only expect habituation and sensitization at the most.

I do not expect that human nature is such that our old habits will be cast away at this one suggestion. Its main purpose is to expose the problem and to shed light on why the nature versus nurture question has generated so much heat in the past. It is true that a single response is a deterministic kind of behavior in the simple sense that it is relatively rigid. It is also true that the origin of culture has been greatly influenced by the rise in importance of multiple choice behavior.

Mixed Single and Multiple Choice Responses

The number of cases where particular behavior is apparently governed by both a single response and a rudimentary kind of multiple choice response is surprising. It is not clear if this is so because one is observing the evolutionary origin of the multiple choice behavior and the remnants of the more primitive single response behavior have not yet disappeared or if it is a case where the selection pressure for the behavior is so strong that nature plays safe by covering the matter both ways. A third, and more interesting possibility,

[2] In the case of *single response* behavior it must be understood that the response may not be absolutely rigid, but could vary in degree, depending upon the intensity of the stimulus, the condition of the responding individual, and other extraneous circumstances. Only one general kind of response is possible in a particular setting. In the case of *multiple choice* behavior more than one general kind of response is possible in the same setting.

is that for different reasons both kinds of response are adaptive for the species; they serve different adaptive functions in the same group of individuals. I shall return to this point after some examples.

An interesting study has been made by S. C. Wecker (1964) on habitat selection in deer mice. This species is widespread over North America and inhabits a large variety of habitats, a reason often given for subdividing it into numerous races or subspecies. Wecker compared the desirability of an open field with that of a woodland habitat by constructing a large series of cages with connecting runways, with half the cages in a field and half in a woods (Figure 37). The runways had a system for recording the passage of the mice, and Wecker used a series of criteria to measure whether they preferred to be under the trees or out in the open. If a mouse from the field was introduced, as one might expect he showed strong preference for the field habitat. A laboratory stock (for a dozen or more generations), on the other hand, when introduced as adults, showed a strong preference for the woods.

The next question was to what degree this could be influenced by where the mouse was brought up. By rearing young from both stocks in isolated pens either in the field or in the woods it was possible to show that the field strain of deer mice retained a fairly strong preference for the field, even after they had been reared in the woods, while the laboratory (wood-preferring) strain showed a strong preference for the field habitat if it had been raised in the field. These results suggest that a mixture of single response behavior (when the field stock prefers the field even though it has been reared in the woods) and multiple choice behavior (when the laboratory stock, which normally prefers the woods, switches to preferring the field habitat if that is where it has been raised). In the latter case the learning must occur at an early stage of development; it is presumably akin to the kind of learning we associate with imprinting.

The question of why mice select their habitats two different ways is intriguing. Wecker suggests that it may be another example of the Baldwin effect. In 1896 J. M. Baldwin, then of Princeton University, put forward the idea that in some instances behavioral changes could precede some more fixed evolutionary change and create a climate favoring the latter. In modern terms, the behavior provided the conditions that permitted the slower genetic fixation

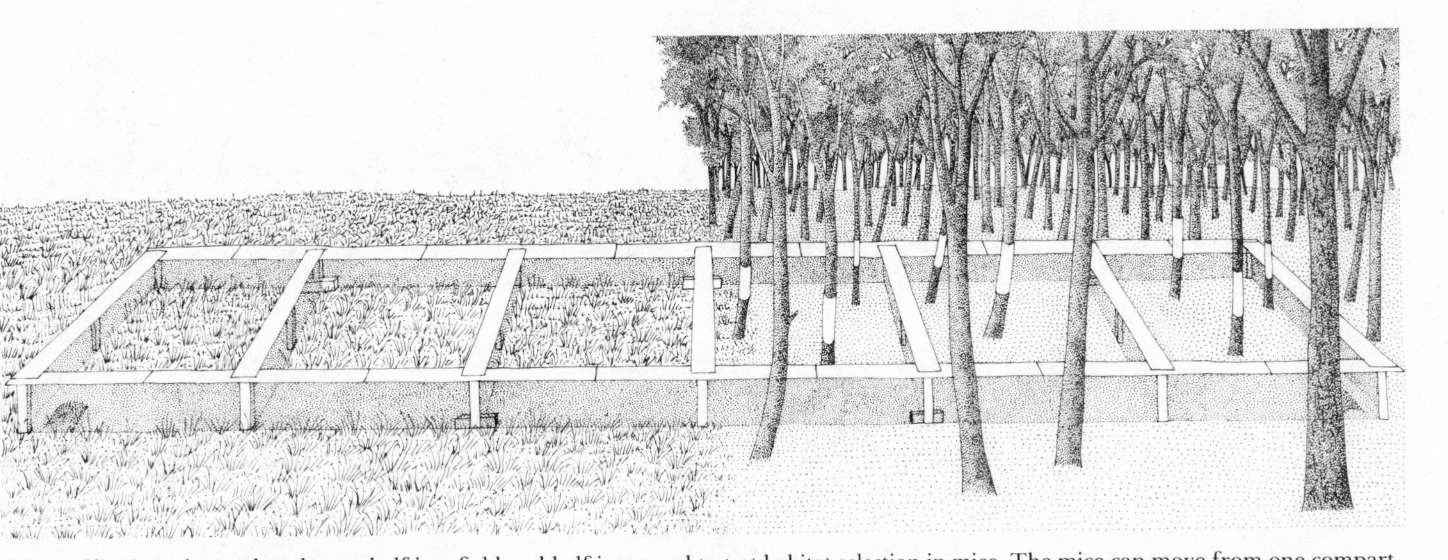

Figure 37. Experimental enclosure half in a field and half in a wood to test habitat selection in mice. The mice can move from one compartment to another and their movements recorded. (The actual enclosure of S. C. Wecker was 100 feet long by 16 feet wide and contained 10 compartments.) (Redrawn from S. C. Wecker 1964.)

of the phenotype.[3] To be more precise, one assumes that a character, such as habitat selection, can be simultaneously genetically fixed and environmentally triggered because of a historical accident. Each method of determination of the character may be the result of selection, but selection at different times in the evolutionary history of the animal. The fact that they both exist at one time is assumed to be because the first to appear was not erased by the appearance of the second.

My colleague Henry Horn has convinced me that it would be at least equally plausible, and perhaps far more so, to assume that for different reasons both the behavioral and the genetic responses, that is, both the multiple and single choice responses are independently adaptive, and that it might be advantageous for an organism to have both. He suggests that when a mouse population is high and the most desirable habitats for a particular strain are crowded, the best strategy would be to move to a less favorable, but uncrowded habitat. To do this, and adapt quickly, it would be helpful to have this shift under behavioral control; to have it so that it can be learned. But if the population again becomes sparse, it would be more effective if the offspring returned to the habitat to which they are best adapted. This is achieved by the genetic component of habitat selection coming to the fore after having been superceded during the period of overpopulation. In this way there might be a simultaneous selection pressure for both methods of habitat selection.[4]

Homeostatic Flexibility

There are a number of phenomena that involve self-regulatory behavior, in the sense that after being disturbed the system returns to equilibrium. A perfect example comes from the well-known researches of C. Richter (1942) who has shown that rats and other

[3] Classical examples of a Baldwin effect involve anatomical features, such as the callosities on the soles of our feet. Abrasion causes the epidermus to thicken, but this thickening is apparent in the uterus before birth. However, callosities are hardly multiple choice responses because the two choices are simply whether to respond or not. For a fuller discussion of earlier views on the Baldwin effect see G. G. Simpson 1953; C. H. Waddington 1957; E. Mayr 1963.

[4] In the case of callosities one might argue that both thickening upon abrasion and preparation for walking by thickening in the embryo are equally adaptive, serving related, but different functions. One helps the functioning of a walking individual; the other protects an infant in her first steps.

mammals, including man, can automatically select a balanced diet. In the simplest sort of experiment a rat will be given a series of pure foods (for example, sucrose, casein, olive oil, salt, water, sources of vitamins), and the self-selected diet will permit optimal weight increase at a remarkably low caloric content. Particularly revealing experiments were performed by removing an endocrine gland: if the adrenals were removed, a rat would eat abnormally large amounts of sodium chloride, and in this way stay alive for an extended period. A rat with the parathyroid glands removed showed a large, but self-sustaining appetite for calcium lactate. This special yearning for calcium lactate was also shown in lactating females. The method is so sensitive that Richter believes that it can be used for diagnosis; an abnormal appetite for some substances will reflect some internal physiological difficulty.

In this case we certainly have multiple choice behavior. It seems to be closely tied into the total physiological needs of the body; in fact, it appears to be one of the prime mechanisms of keeping the internal homeostasis necessary for survival. A deficiency in any one of the basic foods or an internal injury producing an imbalance that needs to be righted are compensated for by behavioral choices.

Some recent studies by various workers have provided a better insight into how the rats can manage to compensate for their food deficiencies.[5] The most important point is that there is a considerable component of quick learning in the process.[6] This discovery stemmed from Richter's work on poisons: a rat will try a new food in very modest amounts, and should the food produce sickness, the rat will no longer touch it, a considerable problem in devising an effective rat poison. One such lesson is enough. A rat with a deficiency, such as the lack of a specific vitamin, will avoid any food lacking the vitamin even when very hungry; it may starve to death if this is its only choice of nutriment. Apparently the rat feels unwell after eating the deficient food and, therefore, puts it in the same category as poison. In order to keep a balanced diet the rat must learn to avoid food that makes it feel sick, and in this way it will tend to favor food that provides all its needs. This quick learning is an efficient mechanism and probably accounts for much, al-

[5] Details can be found in an excellent review by P. Rozin (1976), one of the contributors to these discoveries.

[6] See p. 120.

though not all, of the rat's ability to maintain a healthy intake of food.

The selective advantage of diet regulation is straightforward: anything that helps to maintain homeostasis and to return the internal physiology to some viable level after a disturbance will obviously be of exceptional significance as an adaptation favored by natural selection.

Advanced Behavioral Flexibility

Let us consider now those cases of behavior where an animal is capable of making a choice. In general the choice is based on past experience, and therefore memory is involved. There is nothing stereotyped in the behavior; it is a true multiple choice response. To use E. Mayr's (1974) terminology, it is an "open" as opposed to a "closed" program of behavior.

Let me begin with an example that, as a developmental biologist, I find especially helpful. M. J. Wells (1 962) has done some interesting studies on the prey-catching behavior of the cuttlefish *Sepia*. This highly developed mollusc has two long tentacle arms normally contracted into the body. If a small shrimp is spotted, the *Sepia* moves to just the right distance and orientation with respect to the prey, and then with surprising rapidity, it shoots out its tentacle arms, grasps the prey, and pulls it into its mouth (Figure 38). The whole procedure, even in the adult, is a fixed action pattern. However, there is one very important difference between a young *Sepia* and a mature one. In the young the prey caught is of one type only and cannot be altered by experiment. A small species of shrimp of a certain size elicits the response and nothing else will. In adults, however, it is possible for the individuals to choose between different foods. Furthermore, if something turns out to be difficult and dangerous to catch, such as a crab, they will learn and remember how to catch it with appropriate caution and skill. One of the changes that occurs between youth and maturity is the further development of the brain, including the formation of the lobe responsible for learning and memory. So the young *Sepia* is physically incapable of multiple choice behavior; it has not progressed farther than single response behavior.

E. Mayr (1974) makes the interesting point that under some circumstances a multiple choice behavior (his open program) is adaptive, while in others it is advantageous to have single action re-

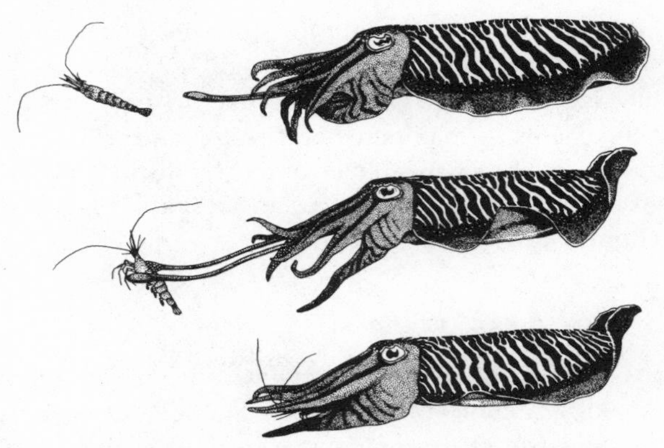

Figure 38. A cuttlefish (*Sepia*) catching a shrimp. Beginning from above, the *Sepia* moves to the correct distance from the prey, rapidly shoots out its long tentacles to catch, and pull it in.

sponses (his closed program). He points out that often the latter is associated with behavior between animals, either of the same species or of different species, while the multiple choice behavior is associated with responses to the overall environment (his "noncommunicative behavior"). Perhaps the most general point is that the greater the unpredictability of the environmental signals, be they from other individuals or the physical environment, the more effective will be a flexible response system. In general, an individual vertebrate will tend to have a rigid and ritualized relation with other individuals of the same species and may have a relatively predictable relation with other species. But the vagaries of the overall environment can be considerable. Let me give some examples to illustrate these points.

Among single action responses, some of the best known are found in courtship behavior. In insects, fish, and birds there are many well-documented cases where there is a series of actions and reactions on the part of the two sexes, each one of which is essential to elicit the next response in the opposite sex. These steps have been analyzed and are so well-known that it is perhaps hardly necessary to illustrate, but let me briefly remind the reader of the classic case of courtship in the three-spined stickleback (Figure 39). There is a fixed sequence of events that can only take one course: the female displays her bulging belly, the male advances and shows

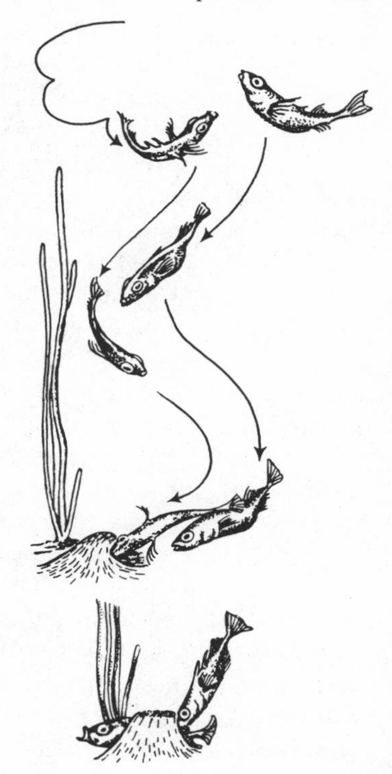

Figure 39. The mating behavior of the three-spined stickleback. Above the male is on the left; below the female is in the nest laying eggs and receiving encouragement from the male. (From N. Tinbergen 1951.)

his red abdomen in return; in a zigzag dance, he courts and turns toward the nest while she follows and enters the nest; he trembles and prods her to stimulate egg laying, after which he fertilizes the egg. It has been shown that if a three-spined and a five-spined stickleback are put together, there are slight differences in the ritual, so it is terminated midway. The same has been shown for interspecific courtship in fruit flies and many other species. In fact, the reasonable assumption is that these differences in courtship behavior between species provide a mechanism to preserve species isolation and prevent hybridization. It is easy to make the further assumption that it is advantageous to have such rituals fixed into single action behavioral responses, for these work as isolating mechanisms by virtue of the fact that any alteration might break down the barriers.

In his discussion, Mayr (1974) cites a most interesting study by J. D. McPhail (1969) that demonstrates clearly how solidly fixed into the genome such a single action response might be. Except for one watershed on the Olympic Peninsula, all male three-spined sticklebacks have a red belly, which is used in both courtship and aggression in territorial defense. The Olympic Peninsula strain males are jet black, probably as a defense mechanism against a predator. It is postulated by Mayr that this subspecies originated in a glacial lake some 8,000 years ago, yet when McPhail presented a female from the Olympic Peninsula with the choice of a black- or a red-bellied male, she chose the red significantly more often than black.[7] Clearly the genome has a long memory; this single action response is indeed rigid and well fixed.

There are other examples of single action or ritual behavior between members of a species, such as aggressive displays, that are as important as courtship displays. Many of the behavior patterns between offspring and parent fall into this category, but by no means all. In fact, this is one area where there is a strong evolutionary trend toward multiple choice behavior. It is clearly adaptive for the young of some species to recognize the parent and the parents the young at an early stage. This involves two major steps: one is to recognize parents as a species, and the other is to recognize one's own parents. The first step is apparently an innate, single action response in some species; one is born recognizing one's own species. An obvious example would be parasitic birds. The brown-headed cowbird recognizes a fellow cowbird even though it was raised by a different species. In other species, the first step is a multiple choice response. Imprinting is the classic example of learning during a critical phase in infancy. This is shown particularly clearly by ducks and geese. The second step is when a bird recognizes its own mother and father. This can be readily observed among colonial birds such as gulls or penguins, where the nests lie close to one another. The ability is apparently learned in the usual sense; the knowledge may be secured sometime in the first five days after hatching and even before hatching, in some instances (B. Tschanz; see Jellis 1977). The adaptive significance of such an ability is thought to arise from the intense competition between neighboring chicks and neighboring adults.

Prey-predator relations that involve different species also involve both single and multiple choice responses. I have already given an example in Wells' study of cuttlefish capturing shrimp and other prey where there is evidence for both kinds of behavior in one life

[7] Thirty-two times out of forty-seven observations ($p = 0.01$).

history. If one looks at the escape responses from predators of other animals, such as birds running to cover at the sight of a hawk, they may be the single action variety. Again, the advantage of its being single is obvious; even a callow chick will have a suitable escape behavior in the face of danger. There are, on the other hand, examples of strategies toward predators that may involve a multiple choice behavior. Many birds may either fly away from a predator or attack it by mobbing. The choice may depend upon many factors, such as the aggressiveness of the predator and the number of fellow birds of the prey species that might be present. Consider a hawk sitting on a perch looking for rodents; if he sees more than one, he has to choose between them. An osprey, as it hovers over the water, must not only decide between fish, but whether the conditions are favorable for a dive. These choices may seem minor, but they are legitimate alternatives that need a decision.

The conditions particularly favorable to multiple choice responses are, according to Mayr (1974), just such factors as food selection and habitat selection discussed in a previous section. He feels that these flexible responses to the larger components in the animal's environment are of special significance because they will permit the animal to invade new habitats more readily and take on new sources of food. It is the kind of flexibility needed for "macroevolutionary" progress. He suggests that many major events in the evolution of the vertebrates should be seen in the light of behavioral responses. As examples he gives the invasion of air by flying insects, pterodactyls, birds, and bats that are thought to be the result of flexibility in locomotory behavior. Again, the evolution of lungs comes from the behavior of fish in stagnant water periodically rising to the surface to gulp air. He says, "those individuals had the greatest evolutionary potential who were able to undergo the most rapid adjustment to changes in the environment or to adopt a new way of life" (1974: 657).

This view reinforces a point I made much earlier in this book, that a pattern of behavior is, as far as natural selection is concerned, no different from any part of the anatomy of an animal. They are both ultimately gene-controlled, although as we have seen, the gene control of behavior may be such that alternate choice behavior is possible. We are not yet talking about Lamarckian or cultural transmission of information; every aspect of the behavior described so far in this chapter is Darwinian in the strict sense of involving gene changes through natural selection. Nevertheless, the flexibility afforded by the multiple choice behavior does permit an animal to find and exploit new habitats, new foods. These new niches will in turn produce new and different selective forces that will change all parts of the genetically controlled phenotype, including not only body structure but also the structure of the nervous system that will result in changes in behavior. Even though this flexibility is totally different from that found in cultural changes, it nevertheless is the kind of flexibility that has produced a foundation for culture. One can think of the appearance of culture as a new niche that arose from the experimentation of animals with multiple choice behavior. And as a macroevolutionary step, it is undoubtedly the biggest of them all.

Alternative Stable Strategies

The last kind of flexibility I wish to discuss is of a totally different sort. It involves behavioral strategies that amount to alternative ways for stability to be achieved. In recent years considerable interest by numerous authors has developed in the question of alternative mating systems and the related problem of parental care.[8] Why is it that some species are monogamous, some polyandrous, some polygynous; why is it that in some species both sexes care for the young, and in others only one of the parents? These and related questions are thought to be directly answered by the forces of selection. The postulated causes concern the behavior patterns, the basic physiology of the species, and the effects of the ecological environment, such as the availability of food, its distribution, and the variability of the climate. But before discussing my main point concerning these strategies, let me briefly describe some of them and show how they vary.[9]

Consider first the physiological costs of reproduction. Many authors have pointed out that in mammals only the females carry milk and therefore have a physiological involvement in offspring rearing that is absent in the male. This freedom for the male is thought to explain why polygyny is so frequent among mammals, but there is a significant number of monogamous species amongst mammals

[8] See, for example, G. H. Orians (1969); R. C. Trivers (1972); S. T. Emlen and L. W. Oring (1977); J. Maynard Smith (1977).

[9] The reader is urged to look at J. Maynard Smith's (1977) paper upon which I have relied heavily, both in the examples presented and, as I shall show, in the ultimate point about these strategies.

also. In the latter cases often both parents will be involved in the care and protection of the young, but presumably because of lactation, there are no known cases where the male exclusively cares for the offspring. As J. Maynard Smith (1977) points out, it is surprising that lactation in males has not evolved in mammals.

It has among birds. The crop secretion of pigeons known as pigeon's milk is produced by both sexes, with the result that they can contribute equally to the care of the offspring. Nevertheless the female does have a greater physiological investment in the form of manufacturing large eggs. In mammals, again the female has this further burden either in the form of eggs (monotremes), the pouch and extended suckling (marsupials), or the period *in utero* (placentals).

On the other hand, in fish the difference in the physiological cost between producing small eggs and large masses of sperm is less obvious. It is not surprising therefore that in many fish, such as the stickleback or the seahorse, the male is solely involved in parental care and the female deserts. The stickleback male builds a nest, defends a territory, and lures the female to the nest. After she deposits her eggs that he fertilizes, he chases her away, and he both cares for the eggs by fanning them with his tail to keep them well aerated and protects them at all times. Note that there is no pair formation and that the young do not need to be fed.

In birds the male becomes the sole parent in quite diverse ways. The male South American rhea, an ostrich-like bird, acquires a harem with as many as fifteen females. Each will lay an egg every other day for seven to ten days in a nest built by the male; when he has accumulated about fifty eggs, he drives off the females and does all the incubation (Figure 40). The rejected female will join the harem of another male and the process will be repeated; some females may do this as many as seven times with different males. At any one moment the mating system would appear polygynous, but it is also polyandrous. The polyandry is a very subtle and clever scheme on the part of the females that even avoids damaging the male's ego.

Another way in which the male is an active parent stems from the interesting phenomenon found in some shore birds that has the curious label "double-clutching." The female will produce two clutches essentially at the same time. One becomes the responsibility of the male and the other her own. Reproductive success is crucial and many of these birds nest in Northern areas where the

Figure 40. A male rhea incubating a full complement of eggs and discouraging the approaches of females eager to lay more.

Figure 41. An American jacaña. (Both sexes have similar plumage.)

breeding season is very short: double-clutching is an efficient way of having more offspring in a very short time. Other birds from more temperate regions commonly have two or more clutches in succession. The extreme example of double-clutching (in fact, it is multiple clutching) is found in the South American jacaña studied by D. A. Jenni and G. Collier (1972) (Figure 41). The female has a large territory and within it she permits a number of males to have subterritories. Each of the latter builds a nest she fills with eggs that they in turn incubate and tend. It is polyandry on a large scale, and as Jenni (1974) points out, it probably evolved from double-clutching.

Instances where both parents appear to be equal partners in domestic chores are common among birds. Many sea birds will share incubation so that the bonded pair alternate, often one trying to push the other off the nest if their broodiness drives are not properly synchronized. The free parent will have a turn to go out to fish at sea, and when there are young, he or she will bring back food to the nest. There is an interesting additional feature in the behavior of some species of penguins, where the nesting ground is a distance from the sea. The problem of bringing food back is apparently sufficiently taxing so that often both parents go, and only a few adult members of the colony take care of the young in one huge crêche (Figure 42). It is the first example of a well-established day care center. Monogamy seems to be the main mating system in animals in which both sexes share in parental care.

Examples where the female takes on the bulk of the offspring rearing duties are common in all groups of vertebrates. Females of a species of cichlid fish of African lakes tend their eggs and then their young by holding them in their mouths. These so-called mouth breeders will then allow the fry to wander off, but will quickly scoop them back should danger appear. In birds there is a range of degrees of female dominance in parental activities. In some instances she does all the brooding and is even fed by the male. In other species such as geese and swans, the male is not involved in brooding and the young feed by themselves, but the male helps in their defense and protection; this produces a fast and long-lasting pair bond between the monogamous parents. The other extremes are those cases where the male only sees the female when mating. This is particularly obvious among those birds with an elaborate male courtship display. In the bower birds the male builds beautiful and complicated bowers solely for mating (Figure 43). More than one female may come to the bower and then wander off to her nest and from then on all the parental responsibilities are hers. A variation of this is found in birds with leks, such as sage grouse of the black grous. The males gather at a site to strut and display their feathers in extraordinary ways (Figure 44) and consequently attract females who arrive, mate, and again wander off to fare for themselves and their young. Leks are also found among mammals; a number of African antelopes show this very clearly.

Figure 42. A crèche of emperor penguin young.

Figure 43. The satin bower bird. The female is in the bower built by the male who is courting on the left.

Figure 44. A lek of sage grouse. The males are displaying and a few females have been attracted to the lek. (By Margaret La Farge. Reprinted by permission of Houghton Mifflin Company from *Watching Birds* by Roger F. Pasquier. Copyright © 1977 by Roger F. Pasquier.)

However, this courtship behavior is not a necessary condition for males that abandon their parental duties. For instance, some monogamous male ducks simply desert their partners after mating and this is extremely common among mammals. The females of deer or elephants will form the social groups that may or may not be broken up into harems during the start of rutting season. In elephants they are not, and the bulls are casual visitors when any female is in oestrus in the group.

The enormous variety in both the mating systems and in the systems of parental care make one wonder why this might be so. A plausible explanation has been offered by J. Maynard Smith (1977) who has looked at the matter in terms of his concept of evolutionarily stable strategies.[10] One of his conclusions is that for any one species with its set of basic characteristics, there could easily be more than one evolutionarily stable strategy. In other words, whether the male or the female take over parental duties, or whether polygyny or polyandry is favored may not, depending upon the circumstances, be important, for either one could be equally stable from an evolutionary point of view.

This conclusion appeals to common sense. Certain things must be achieved: the mates must come together for fertilization, and the offspring must be raised and often fed and protected. It could make little difference which partner does what. There are, how-

[10] See Dawkins (1976) for a good discussion of the concept (pp. 74ff.).

ever, certain restrictions. For instance, milk in mammals for some reason is confined to the female as is egg or embryo cultivation in mammals and birds. Since successful reproduction is important to individuals of both sexes to perpetuate their genes, one would expect each sex to expend roughly the same effort in the whole process of reproduction: if the female makes one kind of an investment, then one would expect the male to contribute in some other way. To maximize reproductive success, if a male deserts immediately after courtship and the female handles all the burdens, the male should presumably mate again. This, however, is not always the case because optimal reproductive success might be achieved without the male making this extra effort. The male can parasitize the female, just as in cowbirds and cuckoos both sexes parasitize other species, and the young are successfully reared. However, if the warm season is very short, as it is in the Arctic, the environment imposes a severe restriction on the time available for reproduction and therefore both parents may have to pitch in and double-clutch. In frigid environments the way parents cope also varies; consider, for instance, the alternative strategy of the penguin where both parents will desert the young for quite extended periods and rely on a crêche for the protection of their offspring. No doubt the kind of food, the method of locomotion, the location of the breeding colony and its relation to the feeding area, and many other factors affect the ultimate system for dividing the labor between the sexes, but alternative equilibrium points of different strategies could quite reasonably be expected to be radically different without significantly affecting reproductive success.[11]

This is the main point; now I would like to apply it to human societies. It does possibly explain why we have such enormous variability in mating systems and parental contributions to child care in different societies, and why we can change so rapidly in any one society, as is presently happening in our own. Again, as long as mating occurs and the children are cared for, it seems very reasonable to expect a large number of different kinds of mating systems and child care strategies. Human sex roles can be switched about with

[11] The same kind of argument for alternative stable strategies has been used by G. C. Williams (1975) for reproductive strategies of animals and plants in general. To give an example, animals or plants can either produce vast quantities of small eggs or seed, each one of which costs little, or a very few large ones that involve a large investment of effort. The total effort for each system may be the same, as may be the final reproductive success; but the strategies are radically different.

amazing ease, and the reproductive success of a population is probably not seriously affected. The only thing we have not been able to find ways of changing by cultural transmission is the *in utero* development of the embryo and lactation in women; these are genetically transmitted characters. But despite this, consider the range of variation in human societies past and present. Besides monogamy there are large numbers of examples of polygyny and a few of polyandry; there is the lessening of the pair bond in recent times in Western society by a higher divorce rate and an increase of couples living together out of wedlock. There is the common situation found in most societies where women have sole control of the children's care, a pattern that is presently altering rapidly, for some men are not only sharing child care, but all of the household duties. With the recent beginning of the emancipation of women a cultural change affects one of the genetically determined aspects of sex roles mentioned above; child-bearing women who work may suppress lactation and feed the baby with a bottle so as to reduce the time away from the professional occupation.[12] In such cases the man will sometimes take on all the parental care and housework. These radical changes we associate with the last few decades are possibly as evolutionarily stable as the earlier methods, just as are the variations one sees in different cultures in different parts of the globe. I mean evolutionarily stable in the sense of J. Maynard Smith, where the strategy is such that the genes of the animals involved remain in the population over many generations.[13] Cultural changes can be especially effective in exploiting new situations, and these new situations may have alternative equilibrium points, each one of which is stable from the evolutionary view. My argument is that one could expect similar equilibria for mating and parental care systems whether or not they are governed by genetic or cultural transmission. It would be valuable to examine human reproductive systems in detail in terms of evolutionary stable equilibria, the same way J. Maynard Smith has done for other animals so as to

[12] The suppression of lactation is an ancient custom, especially for financially privileged women who had their babies fed by wet nurses. Here I am only referring to the practice where it permits women to assume another role in society. The effect of this on fertility and the resulting changing patterns of fertility that have resulted from recent cultural trends are admirably discussed in a paper by R. V. Short (1976).

[13] This does not mean that they are necessarily culturally stable and is an entirely different matter quite outside the scope of this book.

see the relationship between the environment and the strategy of reproduction.

Everything in this and the preceding chapters has been a foundation for the last chapter, where we shall examine culture in animal societies. We have skirted close to the subject frequently and even begun the discussion in the last section, but now we can totally immerse ourselves in it. The lesson from this chapter is that there has been a series of evolutionary trends toward flexibility and that these trends are crucial in the origins of culture.

CHAPTER 8

The Evolution of Culture

I have defined culture as the transfer of information by behavioral means to emphasize its difference from genetical change that involves the transmission of information by genes. The time has now come to examine the appearance of culture in different degrees of complexity and sophistication during the course of the evolution of animals. But before beginning this discussion, it is important to review what has already been said concerning the biological foundations of culture. In the previous chapters I developed various themes bearing on the underlying biological properties needed for the appearance of culture, but I have so far only briefly alluded to the relation of one to another. Here I should like to review and summarize these biological bases so that we can see their logical order and relations. In this way I shall try to smooth off and ready a pedestal to support a discussion of the evolution of culture itself.

Summary of the Biological Bases of Culture

The grand sweep of organic evolution shows a series of branch points where the nature of the adaptations splits into two or more directions. Branching is the opening of new niches, and the organism in the new niche no longer competes or competes less with those that remain in the old niche. For instance, in the conquest of land numerous groups of animals and plants came to such a branch point: some forms remained aquatic, while others mastered the physiological innovations needed for a terrestrial existence. The examples of this kind are numerous; here I shall concentrate on those major branch points ultimately leading to culture.

The first one is the separation between the fast and the slow response. It easily gains first place because it is found even in bacteria. In the case of motility, as was stressed earlier, both the motile and nonmotile conditions can effectively coexist and complement one another, and this is true from bacteria to higher animals and plants. Not only can a motile organism, even a single-celled bacterium, respond to environmental conditions and changes in those conditions, but it can do so rapidly.

The second major branch point comes as an extension of the first. In all organisms the genes contain the information necessary for the construction of the survival machines. In motile animals an additional way of storing and processing information was invented in the form of a nervous system with a centralized brain. This not only makes for more effective and controlled movement, but animals with brains can respond to a greater number of environmental cues. Again, this is a true branch point because all those motile animals without a centralized nervous system continue to exist and peacefully inhabit separate niches that do not necessarily overlap with their more advanced and brainy distant relatives.

There comes a point when the brain surpasses the genome as an information processing machine in the amount of information it can handle and store in the form of memory. It is impossible to estimate with any confidence when this happened during early earth history, but certainly at some moment in the evolution of vertebrates the brain became sufficiently large so that it could process and store more information than the genome.

Again this has not meant the disappearance of animals with lesser brains, and certainly there are many invertebrates, fish, amphibians, reptiles, and birds on the earth happily coexisting with mammals at this very moment. The by-passing of the genome by the brain in this way is a very different kind of branch point than the others we have given. As before, the two branches are adaptive, but the change has been a smooth, gradual shift, not a sudden dichotomy as in the other two branchings. However it came into being, it nevertheless amounts to a turning point in the evolution of some animals, a turning point of important implications for the ultimate ability to have complex culture. Four direct consequences of this progressive rise of the brain are involved.

One of these is that the brain, with its systems of managing communication between individuals of the same species, has made integration into social groups possible, and social animals, more than solitary ones, foster culture. This is so simply because culture involves communication between individuals of the same species, and therefore culture and society go hand in hand.

The second and third products of the brain important for culture are the ability to learn and to teach. These are the basic skills that pass nongenetic information from one individual to another. The more primitive ability to learn is presumed to be roughly reflected in the increase in brain size, as is indicated on Figure 45. (In this graph I have also indicated other features of evolutionary change, including the rise of motility, so that it will be possible to follow on the same figure the biological prerequisites of culture, and some attributes of culture itself.) Teaching, on the other hand, is a more sophisticated skill and was developed later in evolution. The richness of language is limited by the degree of the sophistication of teaching, so the hypothetical teaching curve can be considered the same as that of the rise of language.

The fourth attribute of the brain that is important for the appearance of culture is its ability to give more flexible responses, responses that involve alternative choices. As we have seen, this is a very fundamental quality that not only can be used to separate genome and brain information transmissions, but can also show variation in the kinds of brain-mediated responses. The ultimate in flexibility occurs when the brain has more than given, alternative choices; by invention it produces new ones that did not exist before, a property so conspicuously important in human cultural history, although, as we shall see, it is also found in lesser animals.

With this brief summary of the points made in earlier chapters, I shall now turn directly to culture itself and begin with its primitive origins.

Early Beginnings of Culture

There is a kind of information transfer that is more rudimentary than culture; it is an intermediate condition between direct gene transmission and behavioral transmission. This is the transmission of information by the egg cytoplasm to the new developing embryo at fertilization. It is an ancient adage that we inherit more from our mothers than our fathers, and this more is made up of the chemical components in the egg; the sperm of all higher animals and most invertebrates has little cytoplasm and only a few accessories to the chromosomes, while the egg is an enormous storehouse of food in the form of yolk and, more importantly, all sorts of nucleic acids and proteins that play a key role in early development. It would be stretching the words greatly to consider this cytoplasmic information as knowledge and tradition, but in one sense that is precisely what it is. Knowledge, after all, can mean no more than information, and tradition means a repetition of following out the instruc-

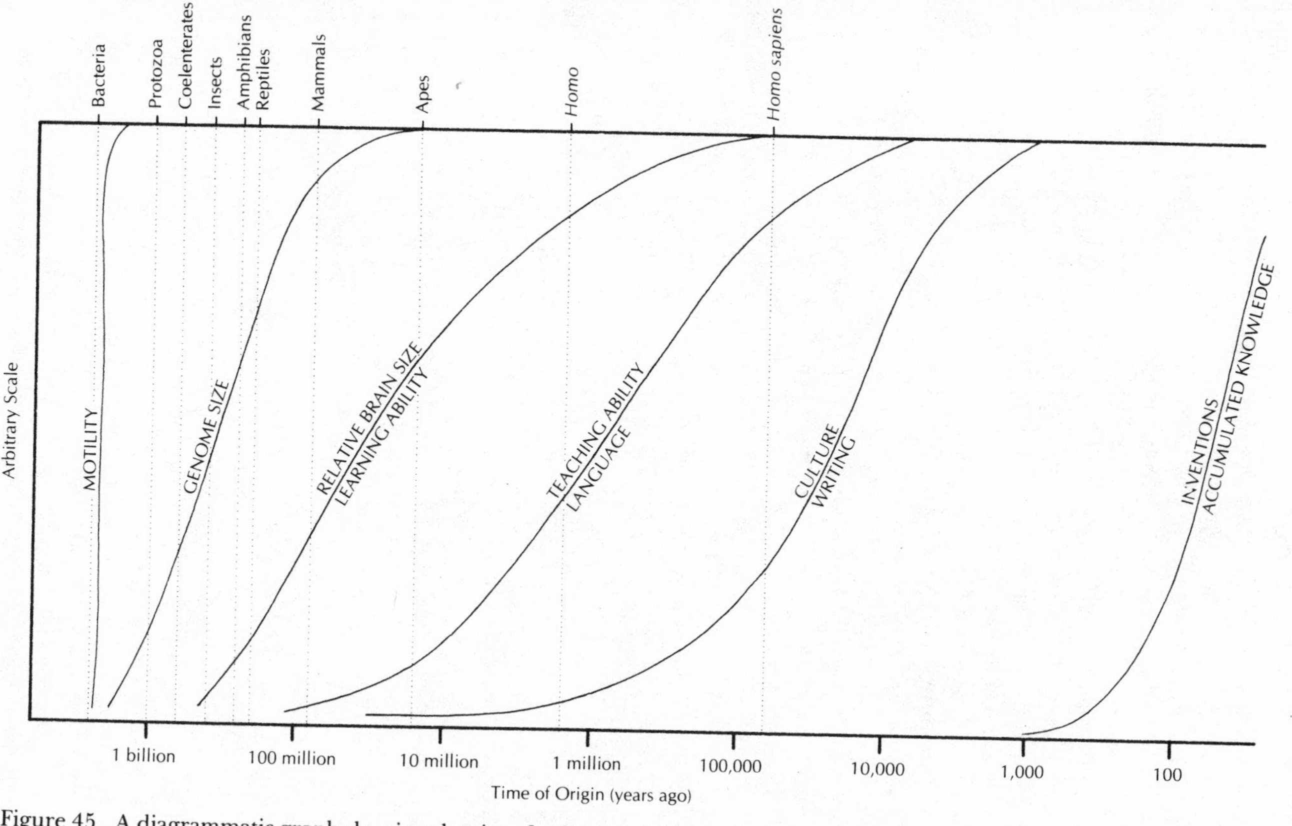

Figure 45. A diagrammatic graph showing the rise of culture and other culture related properties over the course of evolution.

tions of the information. In this case those instructions are faithfully followed in each generation, for instance, in the details of timing and morphology of the cleavage pattern during the early embryogenesis. Another objection might be that the cytoplasmic molecules are gene products from a previous life cycle, but also the behavioral patterns are controlled by the genes in various ways. Why then is this not a perfect example of primitive culture?

The important distinction between true culture and the egg cytoplasm lies in the kind of information transmitted. So far as we know, all the cytoplasmic information pertains to the construction of the embryo. Indeed such instructions have an obvious fundamental effect on the final form of the adult, and therefore on its behavior, but that is only the direct consequence of a normal development. Culture, on the other hand, is the transmission of behavioral information, that is, information which directly affects the behavior of another animal. If that information or knowledge is then repeated in some consistent way, it becomes a behavioral tradition. It is clear that not all nongenetic information is automatically cultural; only the brain can transmit cultural information.

The next question we must ask is what constitutes the simplest brain-mediated culture. Again, as will be evident, the examples still do not fully meet the definition of culture, but they come closer. All the examples, of which I shall give only two, center around the kind of behavior characterized in Chapter Six as single response behavior. That is, the animal had only one possible response to a stimulus (as compared to an animal that had a multiple choice of responses).

Social insects often leave clues in the form of a chemical signal for their nest mates. One such clue that has been studied in some detail is odor trails in ants (Wilson 1971). These are used by one ant to guide others, and they may lead nest mates to a source of food. If a drop of sugar solution is placed near a nest, a worker will return to the nest marking its route with a trail substance it secretes from an abdominal gland. Wilson (1971: 250) showed that the fire ant distributes the trail substance with its stinger: "at frequent intervals the sting is extruded, and its tip drawn lightly over the ground surface, much as a pen is used to ink a thin line." Other ants from the nest are excited by the odor, and once they find the trail, they rush away from the nest and soon find the food. The trail itself does not have a polarity, any more than the trail of a fox can tell the hounds whether the fox is coming or going. Other cues, such as the relation of the trail to the nest, are used to follow the trail in a profitable direction.

This is a single action response to a signal because the signal stimulates only a response of tracking or following. Even though the odor trail itself is chemical, both its deposit and the response to its deposit are behavioral. The limitation of this example as culture comes from the brevity of the accumulation. The odor trail is volatile, and it soon disappears. It is true that it may last long enough to elicit the response from a number of ants, and furthermore those recruits will, in turn, give off more trail substance as they return to the nest; so that at least as long as the food lasts, there is a transmission of behavioral information that accumulates into knowledge and tradition. But after a few hours when the food is gone, the knowledge comes to an abrupt end. Its deficiency as culture is that even as a sequence of messages it is relatively short-lived.

This is not always the case. Let me give another example among insects that is truly remarkable in that not only does the knowledge last over a far greater time span, but the message comes close to having been left in a written form, or at least in the form of an artifact transmitted from one individual to another. Some years ago K. W. Cooper (1957) showed that in a species of wasp that bores into the pith of stems of trees and shrubs, the mother leaves a message for the larva. The brood chambers consist of tubes that penetrate into the stem, and in each tube the female lays a series of eggs starting at the bottom (Figure 46). She adds food with each egg and seals off the chamber with a mud cross-wall, so that the nest consists of a linear series of single larva chambers. After pupation the adults eat through the mud wall and emerge from the opening; but in order to emerge correctly they must be pointing in the right direction. The larvae do this by using the partitions as a cue. The female has so arranged them that the partitions are curved and differ in texture on each side. The larva, by a single action response, orients itself so that its head end is in the appropriate direction.[1] To demonstrate that the partitions do indeed impart the information, Cooper reversed the partitions in a nest and all the emerging young

[1] It must be understood here and elsewhere that by response I really mean the end result. We do not know how the larva responds in detail to the pattern of the partition; it could have a whole series of responses that together cause the larva to orient correctly. It is *single action* in the sense that particular properties of the partition always produce the same larva orientation.

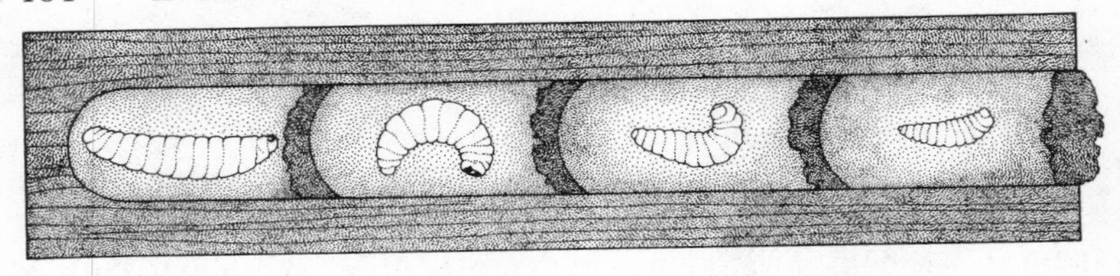

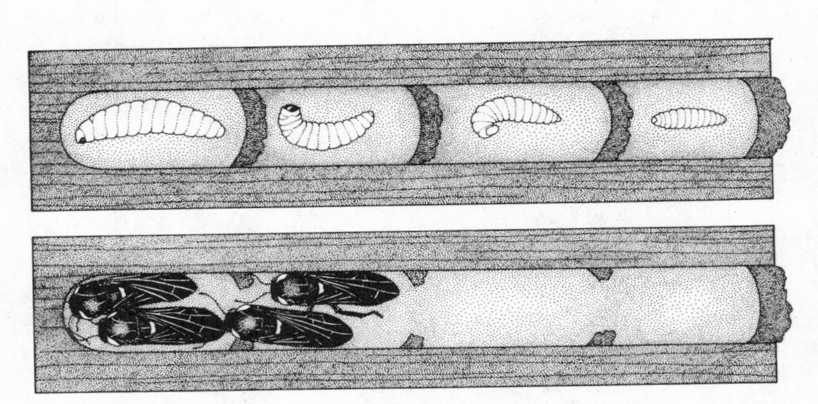

Figure 46. K. W. Cooper's experiment with mason wasp (*Monobia quadridens*) nests. Top: position of larvae and pupae in normal nest made in an artificial hole in wood. Middle and bottom: in this case the partitions have been artificially reversed resulting in a reverse orientation of the pupating larvae (middle) and their death upon emergence (bottom). The food placed by the mother wasp in the cells and the debris that accumulates during growth and pupation are not shown.

adults died because they were pointed the wrong way and crawled into the dead end of the tube, from which they were unable to escape. In a single action response the female laid the cross-walls with the correct polar orientation, and the larvae responded to the texture-convexity by a further single action response and in turn oriented themselves in the correct direction.

This is a far more sophisticated kind of behavioral transmission than odor trails; it not only passes from one generation to the next, which comes close to a tradition, but the information is inscribed on an artifact read by the next generation. The only reason we would consider this a weak form of cultural transmission is that both behavioral steps involve single action responses. This leaves no play for choice or invention. It means therefore that true culture must not only have behavioral transmission, but also must provide and respond to that information in a multiple choice behavioral response. We have, in the last few pages, made our definition of culture more explicit by considering a variety of cases that have some, but not all, of the elements of true culture.

Nonhuman Culture

Many birds and mammals have a primitive form of culture in which there is a transmission of behavioral information and there is a maintenance of accumulation in the form of tradition. Furthermore, the behavior of both the innovators and the followers of the pattern of behavior is very much multiple choice. The only difference between their culture and that of man is that the transmission of the information is by imitation rather than by true teaching which, in any sophisticated or advanced form, is a special gift of human beings.

In his *Descent of Man*, Darwin (1874: 88) clearly recognized the importance of the role of imitation in this kind of nonhuman information transmission. He says, "Dureau de la Malle gives an account of a dog reared by a cat, who learned to imitate the well-known action of a cat licking her paws, and thus washing her ears and face; this was also witnessed by the celebrated naturalist Audouin. I have received several confirmatory accounts; in one of these, a dog had not been suckled by a cat, but had been brought up with one, together with the kittens, and had thus acquired the above habit, which he ever after practiced during his life of thirteen years. . . . A correspondent assures me that a cat in his house used to put her paws into jugs of milk having too narrow a mouth for her head. A kitten of this cat soon learned the same trick, and practiced it ever afterwards, whenever there was an opportunity." If we had been able to follow the kitten and her descendents, and if this method of removing milk from a narrow-necked jug persisted, we would have an excellent example of nonhuman culture.

There are many instances of such behavior among animals, and rather than give a comprehensive list, let me choose a few selected examples to illustrate different kinds of behavior leading to nonhuman culture. I shall make five very rough categories that involve (1) physical dexterity, (2) relations with other species, (3) auditory communication within a species, (4) geographic location, and (5) inventions or innovations.

(1) The examples given by Darwin fit admirably under the category of manual dexterity. To some degree mouse catching in cats is also thought to be passed on by imitation, although clearly there is a strong innate component. The use of tools by many animals provides even better examples: for instance, the holding of thorns in the beaks of Galapagos finches to remove grubs from trees; the use of rocks to break abalone and other mollusc shells by sea otters, and numerous others (carefully catalogued by Wilson 1975: 172ff). One which has been studied with particular care is that of "termiting" by chimpanzees (Goodall 1964). In its natural environment a chimpanzee will take a twig, poke it down one of the openings of a termite nest, twiddle it the right way, and then raise it to lick off the termites (Figure 47). This requires a number of skills and especially knowledge about where to poke, and what kind of stick is best, and how to twiddle the stick. These accomplishments are clearly learned by imitation and they are passed down as a cultural tradition. An interesting parallel occurred in the Yerkes laboratory where one chimp learned to use a drinking fountain and passed this trick on to others. Again the new skill spread rapidly, and remained in the colony, wholly as the result of transmission by imitation.

Another example of exceptional interest is the feeding of the young by oystercatchers. In some localities this shorebird feeds on the mussel, a bivalve with a hard shell that is difficult to crack open. M. N. Norton-Griffiths (1967, 1969) has shown that oystercatchers use one of two methods to get the flesh of the mussel: they either put the mussel on some hard sand and peck or hammer very hard with their bill to break the shell at its weakest spot, or they insert their bill into the open siphon (when the mussel is under water) and snip the adductor muscle which keeps the valves of the mussel closed (Figure 48). Both of these methods are technically difficult and require considerable skill. This is indirectly borne out by the fact that it takes a very long time for the young to learn to open mussels on their own. If the oystercatchers are living in a region where they subsist on worms and other prey that are easy to eat, the young will stay with their parents for six to seven weeks; on the other hand if they are feeding solely on mussels and crabs, the period will last from eighteen to twenty-six weeks. There is some indication that both methods of opening the mussels may be innate behavior; the appropriate movements are seen in all young birds at a very early age. Especially interesting is Norton-Griffiths' observation that an adult bird will either be a hammerer or stabber; no one

Figure 47. Chimpanzees termiting with young watching. (After photographs of J. Goodall 1967, *My Friends the Wild Chimpanzees*, Washington, D. C., National Geographic Society, pp. 49, 51.)

Figure 48. The two techniques of oystercatchers feeding on mussels. Top: hammering; bottom: stabbing. In the former the shell is broken; in the latter the beak penetrates the open siphon (underwater) and the shell is pried open, cutting the adductor muscle. (Based on diagrams of M. Norton-Griffiths 1968, "The Feeding Behavior of the Oyster-catcher [Haematopus ostralegusl]," Ph.D. thesis, University of Ox-ford.)

bird was seen using both techniques during the same period. By switching the eggs from nests of stabbers to nests of hammerers and vice versa, he showed conclusively that the method was learned and not inherited. The unanswered questions concern the degree to which any one bird remains fixed with one technique for life and whether or not stabbers will mate with hammerers. This latter question is particularly interesting since Norton-Griffiths observed that generally both members of a pair appear to use the same mussel-opening method.

Despite these important questions, it is evident that the young birds need a long period of care if their parents are mussel eaters. During this period the chicks do not seem to be practicing; they spend most of their time begging. One presumes that nevertheless they must need that long period, either for the development of their own coordination or for detailed and protracted observations of the technique. What part of it is genetic and what part learning? Are chicks ever reared by parents using different feeding strategies? Very possibly the ability to feed either way is inherited (hence the observation that both types of movement are seen in the young), and the tradition calls forth one to the exclusion of the other.

(2) Cultural transmission of information affecting relations between species largely has to do with the avoidance of predators. There are many examples, of which perhaps the best is that occurring on the Galapagos islands where the birds, especially the top predators, and many other animals show absolutely no fear of man. One can walk right up and touch a bird without its moving. The same behavior may be seen in the marine iguanas and the Galapagos penguins when they are approached on land. But as I. Eibl-Eibesfeldt (1960, 1961) has pointed out, those same amphibious animals become quite terrified if they are swimming and a man joins them in the water. He believes that the answer is that these species have no land predators, but there are sharks in the sea and therefore they have lost their fright reaction on land. Possibly this behavior, at least in its initial stages of development, is not genetically determined; for many species can show such a change in behavior during the period of one life-span. It might be selectively advantageous also for it to become genetically fixed. Darwin assumed that this lack of selection for man as a predator was the reason for the tameness of many of the species on the archipelago.

An interesting study of genetic and learned components of tameness was made by E. M. Adler (1975) with the bank vole and a race of that vole living on the island of Skomer off West Wales (the Skomer vole). The mainland variety is fierce and the young do not allow themselves to be easily handled, while the young of the Skomer voles are docile. Hybrids between the two species give rise invariably to fierce young, clearly including a strong genetic component in which the fierceness of the mainland variety is dominant. There is also evidence that the extent of the aggressiveness or docility can be influenced by the mother during the period of rearing. By exchanging litters between mothers of each race Adler has shown that raising the docile Skomer young with an active mainland mother produced active and aggressive individuals, but in the reciprocal experiment, the Skomer mothers were unable to persuade their mainland offspring to become docile. Therefore high activity can either be directly inherited or passed on in a cultural fashion. From this Adler suggests that perhaps the isolated Skomer island voles have the more primitive "relic" condition and that on the mainland at some subsequent time high aggressive activity could have spread through a population both by rapid cultural change and by slower genetic change. One assumes that the changing predator population on the mainland selected for both of these methods of transmission, but that on Skomer, these predator pressures did not exist.

An excellent example of loss of tameness is reported by I. Douglas-Hamilton (1975) in African elephants. Individuals and groups from areas where they have been shot at relatively little or not at all over the span of the colonial period are comparatively tame, while those from areas where the hunting was intense are shy and dangerous. In the study area of Douglas-Hamilton various groups of females differed: some were docile and others quite ferocious. Furthermore they retained these characteristics roughly to the same degree for a period of a few years, even though he was a constant observer in their midst. He tells of a particularly revealing case of cultural transmission of this sort from the South African Park of Addo: "Here, in 1919, at the request of neighboring citrus farmers, an attempt was made to annihilate a small population of about 140 elephants. A well-known hunter named Pretorius was given the job. Unlike Ian Parker's teams, whose rapid semi-automatic fire liquidated entire family units, Pretorius killed elephants one by one. Each time survivors remained who at the sound of a shot had witnessed one of their family unit members collapsing dead or in its death agonies. . . . Within a space of a year there were only 16 to

30 animals left alive. It seemed that one final push would rid the farmers of their enemies, but by then the remaining elephants had become extremely wary and never came out of the thickest bush until after dark. . . . Pretorius eventually admitted himself beaten and in 1930 the Addo elephants were granted a sanctuary of some 8,000 acres of scrubby hillside. The behavior of these survivors has changed very little, though they have been contained by a fence and are not shot at any more. Even today they remain mainly nocturnal and respond extremely aggressively to any human presence. They are reported to be among the most dangerous elephants in Africa. Few if any of those shot at in 1919 can still be alive, so it seems that their defensive behavior has been transmitted to their offspring, now adult, and even to calves of the third and fourth generation, not one of which has suffered attack from man" (1975: 254-255).

A final example of cultural transmission involving the relation of a species toward its predators is a recent experiment with mobbing in European blackbirds by E. Curio and co-workers (1978). They put two cages on the opposite sides of a hallway and between the cages they installed a cardboard box with four chambers at right angles from one another. By rotating the box 90°, the two birds in the opposite cages either saw an empty box, or with another 90° turn they saw stuffed birds. But first the stuffed birds were not the same: one was an owl that elicited mobbing behavior in the teacher blackbird, and the other was an innocuous Australian honeyeater that the pupil mobbed only because it saw the other blackbird in the act of mobbing. (The pupil could not see the owl, only the honeyeater, toward which it had previously shown very little interest.) The pupil from this first test was then placed opposite a naive blackbird, and when they were both shown a honeyeater, the experienced bird taught the naive one to mob. By repeating this second experiment with six blackbirds in all, they were able to make the birds pass the mob-the-honeyeater tradition from one to another with no loss of vigor in each round of teaching and learning.

One presumes that it is advantageous for an animal to learn rather than inherit information concerning the nature of a predator. The reason could be that in this way an animal can quickly identify new predators and can accommodate to different predators in different geographic regions. Furthermore, by the use of cultural transmission, they may successfully identify a number of predators, something difficult to arrange by genetically determined behavior patterns for each source of danger. Curio and his colleagues suggest that their experiments support the idea that mobbing might be a method of teaching other birds to identify their enemies. If that is correct, then this would be an excellent example of nonhuman teaching.

(3) As I have discussed before, some birds can learn dialects of their songs from other birds. Usually this dialect learning takes place while the bird is young and hears its father or a neighboring male, but as we shall see presently the song dialects are not always learned in this fashion. That these songs are cultural and are passed on from one individual to another is beyond dispute. Unfortunately what is by no means so obvious is the selective advantage of such a mechanism. There have been numerous suggestions, and probably a number of them are correct, for not only do different species show radical differences in their song behavior, but song could have more than one adaptive advantage for any one species.

In the first place, as said earlier, any learned bird song can play a role in individual recognition, and this in turn could be selectively advantageous for a variety of reasons. Another suggestion, made especially by F. Nottebohm (1972; and Nottebohm and Selander 1972), is that dialects permit a group cohesiveness and would tend to isolate those groups into separate geographic regions. In this case isolation is first achieved by behavior and this makes it possible for gene differences to occur in the genetically isolated populations.

This is a reasonable possibility for some species, but in a most interesting study of the New Zealand saddleback P. F. Jenkins (1978) has shown that they are clearly different. These birds are medium-sized passerines characterized by very weak powers of flight (Figure 49). They became almost extinct and in an effort to bring back their numbers, a few individuals were put on some small islands off the coast. The importance of this study is that Jenkins was able to observe marked, individual birds on an entire island over a span of five years. Saddlebacks are unusual in that they do not learn song solely as young, but can learn at any time. As in the chingolo sparrows of Nottebohm and Selander (1972) they do form dialect districts, but there is a big difference. The males are known to wander, and when they arrive in a new area with a dialect different from their own, they quickly learn the new one and adopt it as theirs. Jenkins suggests that the particular adaptive function of this behavior might be to favor outbreeding. In these examples of bird dialects we are still on shaky ground in our guesses as to their adap-

Figure 49. The New Zealand saddleback. (After Bruce Campbell 1974, *The Dictionary of Birds in Color*, New York, Viking Press, p. 219.)

tive significance, but that there are certainly numerous examples of the phenomenon, and in each one the evidence for cultural rather than genetic transmission of the song information is manifest.

(4) One of the most widely cited cases of nonhuman culture is the traditional use of specific routes for migrating birds. As we discussed earlier, day length and hormones have a profound effect on both the urge to migrate and the direction (that is, North in the spring and South in the fall). While these directions may be built into the brain and can be influenced by hormones; in many species the exact path or flyway of the migration or the exact location of the final destination is clearly cultural. Because ethologists have been interested largely in how birds that do not use landmarks as cues navigate, experimental evidence is insufficient on the use of traditional migration routes. To confuse the issue further, in many birds, such as numerous species of passerines, petrels, and shearwaters, the young and adults clearly migrate separately; there is no question of the experienced older birds leading the young.[2]

Nevertheless, in the older literature numerous instances are given where overlapping generations fly together, and it is presumed they take the same pathway each year. Geese are often cited as an example, and it is well-known they have strongly established family groups in which the young and the parents will remain to-

[2] There is always the possibility that in some cases the older birds lead the younger ones for part of the journey.

gether for at least a year and often longer; this includes the first migration to the wintering grounds.

Fortunately at the moment an important series of studies by F. Cooke and his associates of Queen's University in Canada on snow geese is in progress.[3] It has been possible to band the geese and follow the movements of individuals over a number of years; as with Jenkins' study of the saddleback, this seems to be the key to success. Cooke has shown that his geese, which nest and are banded in the Hudson Bay region, go as a family unit to their Texas wintering grounds, but when they return the bonds between parents and offspring gradually begin to fade. Eventually they migrate separately, but so far there has been no accurate way of determining whether or not they follow the precise route they learned in their childhood. The question then of whether the migration route itself is culturally transmitted remains moot.

However, Healey, Cooke, and Colgan (1979) have recently been able to demonstrate that the return to the specific location of the nesting site is inherited culturally. There are two distinct brood areas in La Pérouse Bay near Churchill (Figure 50). Over ninety percent of the year-old goslings return to their natal nesting area and over eighty percent of the banded adults do the same. The tradition wanes only after four years. Furthermore, in the earlier years those individuals that switch from the East to the West areas, or vice versa, are usually ones reared in a region closest to the neighboring area. There is no way this specific choice of an area in the breeding grounds could be genetically determined; it must be passed on from one generation to the next by learning. Therefore some bird flyways might be culturally acquired, but we have much better evidence for cultural transmission in the case of the annual return of birds to specific sites.

K. Lorenz (1977: 158ff.) has an interesting discussion of the phenomenon in geese. He showed that they are loath to go places they have not visited before, and the only way he could induce them to explore new places was to take them there himself. This was possible because he was imprinted upon their brains as a "human foster mother." One day he purposely failed to take them on their regular

[3] For references to the earlier work, see F. Cooke, C. D. MacInnes, and J. P. Prevett (1975). Much of the work of this group is related to the fact that the snow goose has a radically different color phase (the blue goose), and the frequency and the geographic distribution of these two color phases have proved to be most interesting subjects of inquiry.

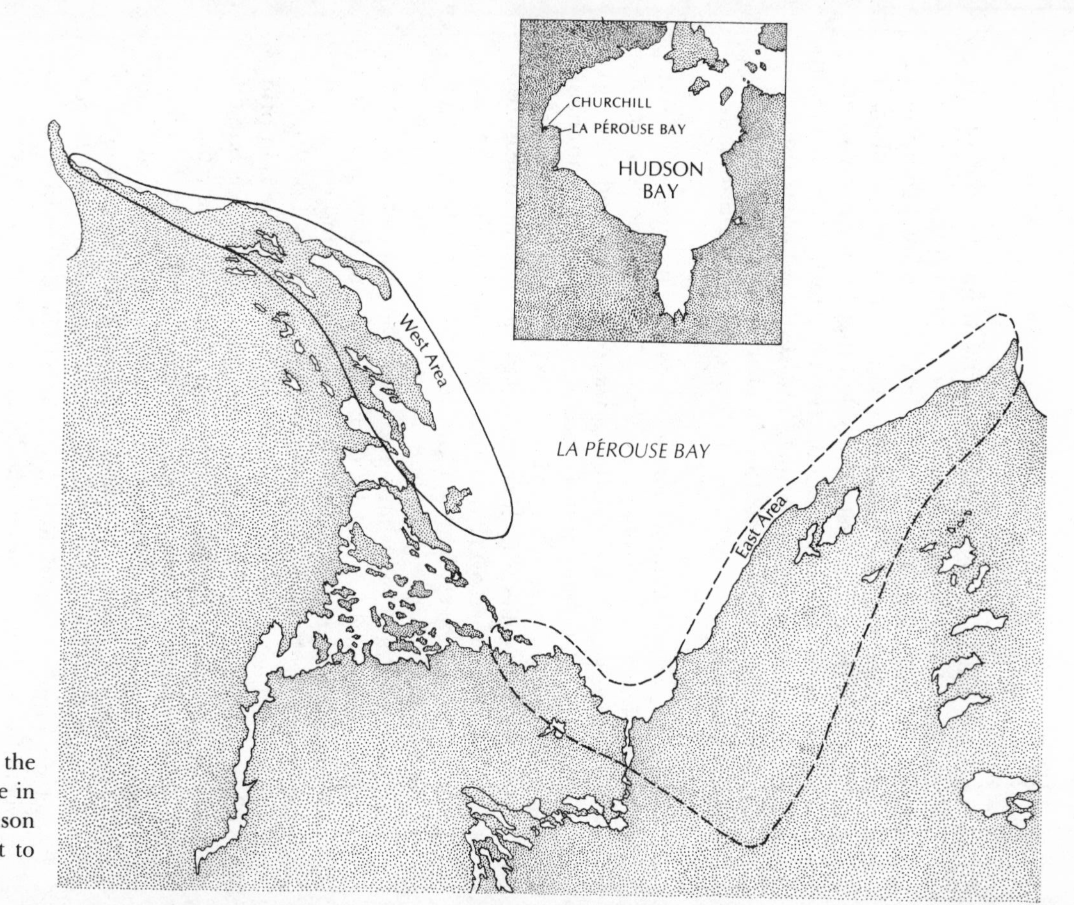

Figure 50. A map showing the two nesting areas of blue geese in La Pérouse Bay in the Hudson Bay. (After a map kindly sent to me by Dr. F. Cooke.)

outing and watched them with binoculars from the roof of his house. After a considerable, noisy wait for him to appear they flew off and systematically looked at all the places he had taken them in the past, starting with the places they visited most frequently and finishing with those least well-known. But they knew exactly where to look and went quite systematically from one to another; they did not try any new places. From these observations Lorenz suggests that they are strongly oriented by familiar localities, and that this tendency is very likely the basis of their conservative traditions in migration.

There are other examples among birds of returning to a fixed location. One of the most striking is the use of a traditional spot for a lek, a place where the male birds will congregate and court to attract the females. Some of these locations have been in continuous use each breeding season for as long as records have been kept.

Monarch butterflies, which also migrate, can live for at least two years, and therefore overlap in their generations. Since they are known to make exceedingly long migratory flights, they may use tradition and follow the same annual routes. More importantly, as in snow geese, they will winter in the same locality each year. For instance, some trees in Pacific Grove, California have been festooned with wintering monarchs for over 70 years, and recently an area of this sort has been found in Mexico (review: Wilson 1975).

This example of butterflies returning to the same site is in principle the same as that of the blue goose although there is one difference. In geese the individuals may return to the same site unaided once they have made a learning run under parental supervision. The young butterflies, on the other hand, always make their journey in the presence of older individuals. However, the older ones must have learned the previous year; they also reach the correct site after one previous learning trip. The only difference between these two examples is that because a goose lives many years there is not an annual, obligatory overlap of generations during migration as there is with a monarch. In both cases geographic information is passed from one generation to the next by behavioral means (imitation and learning) and therefore meets our definition of cultural transmission. That insects can store geographic information by memory is well-known for bees and admirably illustrated in an experiment of Tinbergen (1951) who showed that digger wasps remember the location of their nest by the arrangement of the shrubbery and other objects around it (Figure 51).

Figure 51. Demonstration that digger wasps use learned visual cues to find their nest. The pine cones have been moved from the nest hole and the wasp returning with food can no longer find the entrance. (After N. Tinbergen 1951.)

(5) In all the categories we have discussed above, the possibility of invention or innovation is not only possible, but clearly present. Let us consider each. In the case of physical dexterity, the first use of any tool, such as a twig to capture termites by chimpanzees, would obviously be an invention. It only needs to be discovered once, and thereafter its use is transmitted by imitation.

One of the most celebrated cases of innovation is the opening of milk bottles by titmice in Britain (Figure 52). The trick of pecking through the aluminum foil caps and helping themselves to the cream at the top of the bottle was apparently first invented in a single location, and the skill spread rapidly over the entire British Isles (R. A. Hinde and J. Fisher 1951). This ingenious invention quickly became an epidemic; the other tits learned by imitation how to have their breakfast cream.

The most remarkable of such examples comes from the work of the Japanese Monkey Center where macaques were isolated in groups on small islands, and differences in the behavior patterns of different island populations arose by cultural evolution (review: Wilson 1975: 170ff.). The greatest achievement is that of Imo, the female genius among the macaques. At the age of two she invented washing the sand off sweet potatoes before eating them, and at a later date she found a way of separating wheat from sand by throw-

Figure 52. Blue tits opening the aluminum foil caps of British milk bottles and taking the cream.

ing the mixture in the water and skimming off the wheat from the surface. These discoveries spread slowly through the colony, although in general the older individuals were the last to acquire the new tricks.

If one looks for innovations in acquiring fear and hostility toward new, dangerous intruders, the points made by Douglas-Hamilton (1975) for elephants show that witnessing one or a few incidents of death by shooting is probably enough to transform a docile family group into a dangerous one. Presumably this would be especially so if the matriarch, the leader of the group, changed in her reaction, for she would be the strongest influence in her family. However, the whole group, who witnessed the danger and the horror, could have individually made the innovation of a new attitude toward men.

In the case of bird song dialects, any alteration of the song by one bird could obviously be passed on. That this in fact occurs was shown unequivocally by Jenkins (1978) in his study of the New Zealand saddleback. He observed that new songs, or variations of old songs "arise variously by change of pitch of a note, repetition of a note, the elision of notes and the combination of parts of other existing songs" (1978: 76). This change was an abrupt event and would then turn into a stable song pattern for a number of years. Young birds would learn it so that it became an important component of the dialect of one district. Jenkins calls this change a "cultural mutation," and it is an ideal example of the kind of innovation we are discussing.

Human Culture

In an evolutionary progression, if one passes from primates to man the amount and the complexity of the culture increases enormously. If we ask what is different about man that makes this possible, the answer lies in the fact that besides possessing the improved ability to make multiple choice responses and to learn, man has also greatly increased the art of true teaching. One human being cannot only instruct another, but can impart a wealth of information. Furthermore, that information can be transmitted by a powerful language, and it has even been possible to develop ways of writing the language so that communication can take place through the means of artifacts. Finally, because of such storage methods, we have been able to accumulate information. This most recent accomplishment has meant a logarithmic increase in the total stored knowledge that includes all the inventions and innovations of the past.

A useful way to picture this progression may be seen in Figure 45. The first curve involving brain activities is the one I have labelled learning ability; it is considered roughly equivalent to the evolution of brain size. A much later development is teaching, and since teaching is inevitably linked to a rise in the complexity of language, I have drawn them both as one curve. The next curve, that only begins to rise in a serious fashion for *Homo sapiens*, is labelled culture. Again the linking of the rise of writing with the rise of culture is a very rough approximation. But in general the degree of civilization, were we able to measure it in some satisfactory way, would roughly parallel the span from simple pictures, to more abstract symbols, and finally to a system of representing all the subtleties of language. The only curve with no signs of flattening off at the moment is the last that shows inventions and accumulated information. However, surely the time will come when at least the rate of accumulation will decrease, although we probably have not reached that moment yet.

With considerable care I have tried to adhere to biological problems and avoid discussing anthropological and sociological ones. Nevertheless, it is implicit in everything I have said in this book that there is a biological basis to culture and that although man has the most elaborate culture, it is no less true in his case. We can see the seeds, the origins, of everything we know about our culture in the distant past. This means that every aspect of our culture can benefit from some understanding of the biology from which it sprang. Nor does this deny the obvious and important fact that many features of our culture are new and do not find a counterpart in more lowly animals. If the term is not used in any mystical sense, I am even quite content to call these new properties emergent, for in a straightforward sense, all the evolutionary changes, both structural and behavioral, have new properties that do not exist at a lower level of complexity. But this is an obvious descriptive statement, a statement of fact, and not put forward here as an explanation or as a philosophical prop of any deep significance or practical value. It simply says that more complex structures may act differently than less complex ones. The danger lies in believing that a hierarchical level takes on a life of its own and thereby loses all connections with its lower level.

Unfortunately battle lines at the juncture between biology and the social sciences have been drawn. No doubt traditions and territoriality and inflammatory statements on both sides make it difficult to communicate peacefully across the border. It seems to me fairly obvious that it is important for the biologist, and more particularly the sociobiologist, to realize that his recent flashes of insight that have come, for instance, through the aegis of kin selection, will not solve all the problems of the social sciences, but may shed some bright light on aspects of human social behavior. The social scientist, on the other hand, must face the possibility of some biological information being extraordinarily useful to him, and certainly it should not be rejected for doctrinaire reasons.

Today, as always, the evolutionary biologist teeters at the edge of a serious difficulty. Because of the properties of natural selection, it is exceedingly hard to prove that any one character appeared as a result of selection. There are now a few well-known cases, of which the most celebrated is H.B.O. Kettlewell's (1973) work on industrial melanism in the British peppered moth. It was shown that the dark phase of the moth in regions near industrial towns increased, and more recently the light phase increased as the pollution has come under control (Bishop and Cook 1975). That the relative frequency of these two genetically different strains is controlled by selection has been shown in experiments in which the eating of the moths by birds (their natural predators) has been scored. When the moths are placed on light or dark tree trunks, the contrasting coloration is not protective for the moths (light moth on a dark tree and vice versa) but is protective for ones that blend (light moth on light tree and dark on dark).[4]

While there are some other well-authenticated examples, they amount to a mere handful; the vast majority of cases are presumed rather than proved. Darwin was criticized (and still is by some) for the very same reason. This was especially so at the turn of the century when it was rather generally thought that natural selection could account for the elimination of undesirable traits and not for the appearance of new ones. For instance, this view is very clearly expressed in D'Arcy Thompson's famous book *On Growth and Form* first published in 1917.

These difficulties become acute when one seeks the effect of natural selection on culture. Here one not only has all the problems of proof just discussed, but the unyielding and uncooperative question of whether or not any particular act of social behavior is genetically or culturally transmitted. In order to sidestep this intractable problem we shall now examine the question in the most general terms: is culture, as a means of nongenetic transmission of information, adaptive? Could this explain why it has come into being?

The Adaptive Advantages of Culture

Before embarking on any kind of speculation (for speculation it must be), let me begin by stating something everyone knows, for I want the sophisticated reader to understand that I do too. There is no reason why every structure, or every color, or every behavior need be adaptive and the fruit of natural selection; there are at least three good reasons why they need not be. To begin, as originally suggested by Sewell Wright (1931), isolated small populations might produce genetic changes by chance. Here we shall confine ourselves to those cases in which one or a few individuals begin a population and therefore their initial genetic constitution will obviously have a big effect on the final gene composition of the population. This has been called the founder principle or effect (Mayr 1963) and the Adam and Eve effect (Maynard Smith 1975).

The second argument for the nonselection of genes and their

[4] Kettlewell (1955) showed that behavior of the moth is also involved and reinforces the selection. That is, dark phase moths seek dark backgrounds and light phase ones light backgrounds in this way making the cryptic coloration even more effective in protecting the moth from the bird predators.

characters is that some might be selectively neutral. M. Kimura and T. Ohta (1971) and others have taken the position that neutral mutations do arise and show themselves, for instance, in the variations of amino acid sequences in proteins. The opposing view is that all characters will inevitably be subject to selection. The problem, it seems to me, can easily be avoided by assuming that there is a continuum between no selection pressure for a gene-controlled trait to one where there is an obvious and sizable pressure. This means that many mutations might not be neutral in any absolute way, but encounter such slight selection that the resulting change of their frequency in a population can be considered negligible.

The third way in which a trait might be untouched by the pressures of natural selection involves a developmental argument. Genes can be, and often are, pleiotropic; that is, the gene will have more than one phenotypic effect. Suppose, for instance, that a gene had two effects on the phenotype, one very good, and one neutral or slightly deleterious. As long as the positive selective pressure for the good phenotypic character was greater than that for the bad trait, the latter would be carried in the population, even though it is undesirable. One could have a single gene mutation resulting in an increase in offspring in female birds (for example, increased clutch size) but at the same time reducing their ability to catch food (for example, they move more slowly because of increased weight). This mutation will be selected for if the increase in offspring is not offset by the disadvantageous effect of inefficient food gathering. But to illustrate how complex these questions might be, we should note that food-gathering efficiency will depend on food availability; if there is plenty of food, the loss of ability to catch it will be less important than if the food is scarce. It is also possible that such a limitation would only occur during the nesting season, and if the male fed the female (as occurs frequently), the disadvantage of the sex-linked gene would be minimized by cooperative behavior on the part of the male.

A related phenomenon that may also confuse the question of whether or not a particular behavior has appeared because of selective forces is the matter of alternative stable states discussed at the end of the last chapter. In our examination of mating systems, it was pointed out that there were alternative and possibly equally stable strategies, for instance, in the mating and parental care systems of birds (Maynard Smith 1977). Male or female or joint tending of the young (and all the variations that we discussed) could be alternate and equally successful ways of doing the same thing. The examples given were all related to reproduction, but one could just as well have considered prey catching. For example, either being able to swim very fast like a mackerel or being able to entice fish into a trap like a sedentary angler fish that hangs out a bait-like protuberance to attract its prey are both effective methods of catching other fish; they are two very different, but presumably equally stable, strategies of food intake. Selection can appear to be operating in opposing directions, but of course this is not the case. Selection is directed to an end result: adequate and effective offspring care or efficient prey catching, and often there is more than one way to achieve these ends satisfactorily.

With all these caveats in the forefront of our minds, let us now consider the adaptive advantages of culture, beginning with the more primitive forms of nonhuman culture. We are not now considering the advantages of quick movement and reactions, an elaborate nervous system, a social existence, or any of those matters we have already discussed. We are talking solely about culture and those features leading quite directly to it.

The crucial point is that certain kinds of information can only be transmitted by behavioral means; to do so through the genome would be so complicated and difficult that it would be impossible. If the transmission of this difficult kind of information is adaptive, then there would be strong selection pressure for culture. Only the nongenetic kind of transmission is feasible, and therefore any gene changes that make this kind of transmission possible will be favored. I intend now to give two examples that illustrate such difficult information that can only be transmitted culturally. One involves the transmission of geographic or topographic information; the other involves the transmission of information that leads to a closer relation between individuals of a species.

Earlier I emphasized the fact that among birds, undoubtedly in mammals, and possibly in insects, returning to the same annual nesting, lek, or wintering sites were instances of nonhuman culture. Let us make the assumption that using a familiar and traditional location is selectively advantageous. In this event it would be virtually impossible for a bird to inherit information about a geographic location it had never seen. Imagine how extraordinarily complex it would be for the genome to devise a method of coding such knowledge. By comparison it would be simplicity itself if the information identifying the location is passed by tradition dependent upon the young birds imitating the older ones. There are many intermediate examples where learning is an easier mechanism for transmitting information than having the whole behavior innate and handled solely by the genome. In effect this means that there will be a whole range of degrees of selection pressure for transmitting information by nongenetic means.

The second example of how a selection pressure for culture might have arisen is more hypothetical, but potentially of far greater importance. It has to do with the interactions between individuals of a species. As one goes up the evolutionary scale of animals, with increasing brain size the ability to recognize individuals of one's own species also increases. Social insects show a very limited capacity to do so; they are much better at recognizing nest mates and foreign individuals are immediately rejected, presumably because they lack the appropriate nest odor.

In birds, pair bonds are often formed, and the mates do learn to recognize one another as individuals. We have already discussed the possibility that some monogamous birds may have developed vocal communication between the sexes for the sole purpose of tightening the bond, largely by making each constantly aware of the other. This is certainly a good possibility in the duetting birds, where they essentially sing one song together. It has been argued that these songs are used to permit the mates to keep track of one another in a dense forest, but this simply provides a selective reason for the birds to make their presence constantly known to each other. It was also pointed out earlier that birds such as mynahs, parrots, and mockingbirds, that imitate song so cleverly, might use this to find unique, but mutually understood sounds that provide a special bond between a mated pair. These bonds between individuals are presumably important for stability in the family unit. In many species of birds the young learn to recognize their parents' call and vice versa, and these bonds are important in the competition for food. Proper food allocation in a nesting colony of sea birds will permit a larger success in producing fledglings, and individual recognition is an effective way of properly distributing the resources.

None of these examples involve any degree of culture, except in the very limited sense that some of the mutually devised calls may be passed on from one individual to another. However, they are the beginnings of the development of an effective way of transmitting information between individuals, which is an essential aspect of culture.

If we turn to mammals, individual recognition and the number of recognition signs and signals between individuals increase. With increased brain size, the signals become increasingly complex, ultimately developing into a genuine language in man. This entire gamut always involves interaction between members of the same species: bonds between mates, between offspring and parent, between siblings, and even between rivals, both related and unrelated. The signals have gone far beyond simply bringing individuals together; it is now possible to have a variety of interaction between them. The information content of the signals grows to the point where it suddenly seems to be staggeringly elaborate and complex in man. Here we are concerned with the adaptive value of each step toward culture and therefore must ask why the elaboration of interindividual signals increased so strikingly in mammals. Undoubtedly the answer lies in the general rules laid down by individual selection. Genes constantly compete in a population. Those genes or combinations of genes that show the greatest fitness (that is, reproductive success) will persist, and favorable new mutant genes will survive in the competition. Therefore, any genetically determined change improving the success in rearing offspring will be favored; the success of gaining food, both as a young and as an adult; the success of finding a mate; in short, all the things that lead to reproductive success. Such changes will include cooperative acts such as pair formation, protection of related individuals, and noncooperative, selfish acts such as taking as much food as an individual can grab or chasing away rival suitors. All these acts, and many others, are essentially competitive. They are the strategies for success of the selfish genes of individuals. Since this is the case, there will be a steady selection pressure for those genes affecting all the variety of cooperative and selfish acts that help individuals and their genes win in the competition. Every other member of a species is a rival; and those that invent, by mutation, genes to improve the chances of reproductive success are clearly the winners. This process involves a strong selective pressure for an increase in the communication between individuals.

But the system of communication itself has spawned the possibility of a new method of transmission bypassing the genes. This means that in order to win in the struggle for reproductive success it might be advantageous to transmit information directly from one individual to another rather than through the genome. Therefore any cooperative or selfish act that is adaptive and that could be achieved by quick signal transmission would be favored over any slower genetic transmission. This bypassing was so successful that the selective pressure on the genes was no longer for more elaborate genetic signals, but for bigger and better brains that could transmit a wide variety of rapid, flexible, innovative signals in a behavioral, rather than in a genetic fashion. This step is the cornerstone of the evolution of culture; and there is every reason to believe it occurred as the consequence of natural selection.

Once the method of bypassing the genome by the transmission of information through brain-mediated actions was established, an extraordinary new era dawned. It was suddenly possible to pass on, accumulate, and even invent new information without any *direct* instructions from the genes. Previous changes that might have taken hundreds or thousands of years could now occur in weeks or days, or even hours. The method was so effective that it did very clearly have adaptive value. The brain information transmitters were successful, and better transmission meant increased fitness. Therefore, intense selection pressure was for the genetically determined biological structure that made the flexible, innovative behavioral transmission possible, namely the brain. A particularly spectacular event was the increase of the brain size in the hominid line. This was achieved by the selection of relatively few genes that extended the duration of the period of growth of the brain. This size increase provided the means for elaborate teaching and the development of language. Language came to the point where it could be represented by symbols, and writing made possible the accumulation of knowledge (Figure 45).

My argument is that culture has been extraordinarily adaptive, and the only way to improve is to select for gene changes that produce a generalized increase in the brain size. There is also an evolution of culture itself, which merges in its most rapid form into history, but this nongenetical evolution can only make special kinds of changes, ones that are circumscribed by the power of the human brain; they cannot change the brain itself. Only genetical evolution can do that.

As before, let me emphasize that even though mammals have been successful in this evolutionary trend, and primates the most successful of all, not all the other vertebrates, the fish, the amphibians, the reptiles, and the birds should become extinct. Selection, more especially in a complex environment, will find many successful solutions each of which correspond to the available ecological niches. This means that the environment cannot only simultaneously support motile and nonmotile organisms, but also ones of different degrees of social existence, different abilities to transmit cultural information, and ones of different brain sizes. This coexistence of a variety of different kinds of adaptations is part of the basic fabric of evolutionary biology.

Let me now pause to emphasize an important point. It s my contention that there were strong selection forces for the process of cultural transmission and that this general pressure provided the impetus behind the spectacular changes that produced a larger brain so rapidly in the evolution of the hominids. The genetic selection was not for any particular set of cultural characteristics, but simply for the ability to transmit information by behavioral means. Clearly the cultures that subsequently arose among those organisms capable of having them must have been of such a nature that they did not adversely affect the rate of reproduction of the species. If such destructive cultural patterns appeared, they would have either disappeared as the population itself dwindled (for example, the Shakers) or the customs would have changed behaviorally by further cultural evolution. Although there are instances where the customs have caused the destruction of a population, reproductive success is evident in the fact that the number of human beings in the entire world has been rising in recent historical times.

Conclusion

At the moment some want to see what acts of human social behavior might have arisen by natural selection during the course of evolution. This is the direct consequence of seeing good reasons to suspect that many equivalent behavior patterns in animals have probably been shaped by selection, by changing the genetic constitution of the animal. The problem in the case of man is that it is especially difficult to demonstrate what components of behavior have a direct genetic basis. Furthermore, many human behavior patterns vary enormously in different societies; and there is every reason to believe that the role of cultural as opposed to genetic transmission of information plays an especially important part.

Alternative stable states can occur in both genetical and cultural evolution, so that the analysis of any one strategy will tell us little about whether the transmission is by genes or memes or a mixture of the two. I have argued here that for both practical and possibly fundamental reasons the more rewarding question of why we have memes at all should be asked. Why did this mode of transmission arise in the first place during the long course of evolution?

My answer has been that culture as a process is by itself of enormous adaptive value. With it, and its rapid system of meme transmission, an animal can perform certain behavioral feats, such as recognition of geographic sites or the recognition of individuals, that would be impossible to achieve by gene transmission. Because of this large positive selection pressure, genes have been selected that have ultimately produced survival machines capable of sending and receiving memes.

We have followed the sequence to this selection process in some detail. The first step lies in the origin of motility, for movement is basic to all organisms that can transmit memes. The next is in the progression from a single response behavior to a multiple choice one. Furthermore the flexibility of these responses increases, that is, the multiplicity of responses, including even that of invention, increases. All this is accompanied by a rise of a communication system involving a greater complexity of the responses to signals, including an ability to learn, and a greater ability to emit signals and teach. Ultimately the signalling system achieved the status of language and writing was invented. Every one of these steps toward meme transmission is assumed to have been adaptively advantageous; at each point in the long evolutionary journey there were reasons for rapidity and flexibility to increase fitness.

The most dramatic step came when memes themselves first became possible in early man. Memes had appeared and are clearly adaptive in other animals, but something happened in primate behavior that differed from all other animal culture by some orders of magnitude. As before, the meme system of transmission continued to provide a competitive advantage, but a new and important genetic change occurred that made the progress in this cultural transmission suddenly increase at an astounding rate. Instead of making the brain larger by totally restructuring it, which would have involved a large amount of time and a vast number of gene changes, there was a change in a few genes that affected the timing of development: the brain was simply allowed to grow for a longer period of time than the rest of the body. By this clever ruse, a small genetic change produced a larger brain, that in turn was masterly at handling memes in a variety of ways including complex teaching.[5]

Animals with large brains do well as survival machines because they adapt to any situation with enormous flexibility and originality. There is undoubtedly some or even much genetically determined behavior as well, but it is difficult to detect. One exception is a presumed genetically determined instinct for survival; it is difficult to imagine any more adaptive behavior pattern, and there appears to be some support or such an innate behavior in the majority of animals with brains.

The instinct for survival is important to culture because a meme in order to be invented or acquired must pass a severe test: If it in any way endangers the lives of the animals concerned, it will automatically be rejected.[6] In other words, once a brain is constructed, it will not thoughtlessly tolerate any meme, but only those that do not adversely affect the survival of the individual. Just as genes are internally selected immediately during the course of development (for any gene producing lethal effects will immediately be eliminated), so also memes are immediately selected by our brains, and those that are suicidal are summarily rejected. This means that many of our customs, traditions, and even moral precepts are handled by memes as well as, if not better than, by genes.

It must be remembered that culture and customs can be considered alternative steady states. This is exactly the point I made earlier concerning mating systems in other animals, and I showed that those same systems can be culturally copied by man with equal success. There is obviously even greater latitude in culturally determined social groups to shift between alternative steady states than in the genetically determined ones of nonhuman animals. No doubt the reason is that if an experimental change devised by the human brain shows signs of becoming disastrous, a further behavioral change can right the situation. In other words we have rapid methods of providing balances and checks that can be (but unfortunately

[5] Obviously many other genetic changes over and above those affecting brain size contribute to the adaptive success of higher vertebrates. Nevertheless, brain size was very likely pivotal in their evolution.

[6] The exception here might be those animals, such as mayflies and Pacific salmon, that have a *programmed death* right after fertilization and the depositing of the eggs. But even in these species, at all other times individuals retain an ardent interest in living. Their death is not a brain-mediated suicide, but rather a genetically determined failure of their body.

are not always) used when things are going in the wrong direction. We do not have to rely on the slow selection of genes.

These conditions for having many different cultures of variable stability and success are the prime interest of the social scientists. They want to know the more delicate forces producing and inhibiting change. They want to know the man-made factors creating the particular conditions of any society and they seek the causes of cultural evolution in the actions of man and in man's customs. They see little reason to worry about what cultural change might in the long run affect his genetic constitution. It is the very fact that the biologist and the social scientist are looking at different causes that makes it difficult for one to see the problems of the other. What biology has to say about culture lies in the genetical evolution of animals and man. The frame is so big, and seemingly so remote, that the social scientist may never consider it. What the social scientist says about any aspect of culture applies to the immediate behavioral acts that brought it into being. In cultural changes there is a great freedom for variation and transmutation that may not seriously upset the biological fitness, that is, the reproductive success of the individuals in any particular civilization. However, the degree of change and variation is not without limits, and the nature of those limits, as well as the causes of change, will continue to be subjects of fascination and importance to us all.

Bibliography

Adler, E. M. 1975. Genetic and maternal influences on docility in the Skomer vole, *Clethrionomys glareolus skomerensis*. *Behav. Biol.* 13:251-255.

Adler, J. 1976. The sensing of chemicals by bacteria. *Sci. Amer.* 234 (4):40-47.

Alexander, R. D., and P. W. Sherman. 1977. Local mate competition and parental investment in social insects. *Science* 196:494-500.

Amadon, D. 1959. The significance of sexual differences in size among birds. *Proc. Amer. Phil. Soc.* 103:531-536.

Aschner, M., and J. Cronin-Kinsh. 1970. Light-oriented locomotion in certain Myxobacter species. *Archiv für Mikrobiologie* 74:308-314.

Baldwin, J. M. 1896. A new factor in evolution. *Amer. Natur.* 30:441-451, 536-553.

Bateson, P.P.G. 1975. Specificity and origins of behavior. *Adv. Studies Behav.* 6:1-20.

Benzer, S. 1973. Genetic dissection of behavior. *Sci. Amer.* 229 (6):24-37.

Berg, H. C. 1975. How bacteria swim. *Sci. Amer.* 233 (2):36-44.

Berg, H. C., and R. A. Anderson. 1973. Bacteria swim by rotating their flagellar filaments. *Nature* 245:380-392.

Bertram, B.C.R. 1975. Social factors influencing reproduction in wild lions. *J. Zool.* 177:463-482.

——. 1976. Kin selection in lions and evolution. In *Growing Points in Ethology*, eds. P.P.G. Bateson and R. A. Hinde, pp. 281-301. Cambridge: Cambridge Univ. Press.

Bishop, J. A., and L. M. Cooke. 1975. Moths, melanism and clean air. *Sci. Amer.* 232 (1):90-99.

Bodmer, W. F., and L. L. Cavalli-Sforza. 1976. *Genetics, Evolution, and Man.* San Francisco: W. H. Freeman.

Bodot, P. 1969. Composition des colonies de termites: Ses fluctuations au cours du temps. *Insectes Sociaux* 16:39-53.

Bonner, J. T. 1965. *Size and Cycle: An Essay on the Structure of Biology*. Princeton, N.J.: Princeton Univ. Press.

——. 1970. The chemical ecology of cells in the soil. In *Chemical Ecology*, eds. E. Sondheimer and J. B. Simeone, pp. 1-19. New York: Academic Press.

——. 1973. Hormones in social amoebae. In *Humoral Control of Growth and Differentiation*, vol. 2, eds. J. Lobue and A. S. Gordon, pp. 81-98. New York: Academic Press.

——. 1974. *On Development; the Biology of Form*. Cambridge, Mass.: Harvard Univ. Press.

Bonner, J. T. 1977. Some aspects of chemotaxis using the cellular slime molds as an example. *Mycologia* 69:443-459.

Brown, J. L. 1975. *The Evolution of Behavior*. New York: Norton.

Campbell, R. D., R. K. Josephson, W. E. Schwab, and N. B. Rushforth. 1976. Excitability of nerve-free hydra. *Nature* 262:388-390.

Capranica, R. R. 1965. *The Evoked Vocal Response of the Bullfrog: A Study of Communication by Sound*. Cambridge, Mass.: M.I.T. Press.

Carpenter, C. R. 1934. A field study of the behavior and social relations of howling monkeys. *Comp. Psych. Monogr.* 10:1-168.

Carpenter, G.D.H. 1949. *Pseudacraea eurytus* (L.) (Lep. Nymphalidae): a study of a polymorphic mimic in various degrees of speciation. *Trans. Roy. Entomol. Soc. London* 100:71-133.

Cavalli-Sforza, L. L., and M. W. Feldman. 1978. Towards a theory of cultural evolution. *Interdisciplinary Sci. Rev.* 3:99-107.

Clark L. R., P. W. Geier, R. D. Hughes, and R. F. Morris. 1967. *The Ecology of Insect Populations in Theory and Practice*. London: Methuen.

Cooke, F., C. D. MacInnes, and J. P. Prevett. 1975. Gene flow between breeding populations of lesser snow geese. *Auk* 92:493-510.

Cooper, K. W. 1957. Biology of Eumenine wasps: V, digital communication in wasps. *J. Exper. Zool.* 134:469-509.

Count, E. W. 1947. Brain and body weight in man: Their antecedents in growth and evolution. *Ann. New York Acad. Sci.* 46:993-1122.

Curio, E., V. Ernst, and W. Vieth. 1978. Cultural transmission of enemy recognition: one function of mobbing. *Science* 202:899-901.

Darling, F. F. 1937. *A Herd of Red Deer*. Oxford: Oxford Univ. Press.

Darwin, C. 1859. *On the Origin of Species*, reprint ed., Cambridge, Mass.: Harvard Univ. Press. 1964.

——. 1874. *The Descent of Man and Selection in Relation to Sex*, 2nd ed. New York: Hurst and Co.

Dawkins, R. 1976. *The Selfish Gene*. Oxford: Oxford Univ. Press.

de Beer, G. R. 1940. *Embryos and Ancestors*. Oxford: Clarendon Press.

Douglas-Hamilton, I., and O. Douglas-Hamilton. 1975. *Among the Elephants*. New York: Viking Press.

Dworkin, M. 1972. The Myxobacteria: New directions in studies of procaryotic development. *Crit. Rev. Microbiol.* 2:435-452.

Eibl-Eibesfeldt, I. 1960. Naturschutzprobleme auf den Galapagos-Inseln. *Acta Trop.* 17:97-137.

——. 1961. *Galapagos, the Noah's Ark of the Pacific*, trans. Alan Houghton Broderick. New York: Doubleday.

Emlen, S. T. 1975. The stellar-orientation system of a migratory bird. *Sci. Amer.* 233 (2):102-111.

Emlen, S. T., and L. W. Oring. 1977. Ecology, sexual selection and the evolution of mating systems. *Science* 197:215-223.

Etter, M. A. 1978. Sahlins and sociobiology. *Amer. Ethnologist* 5:160-169.

Fagan, R. M. 1978. *Animal Play Behavior* (Book in preparation).

Fankhauser, G. 1945. The effect of changes in chromosome number on amphibian development. *Quart. Rev. Biol.* 20:20-78.

Fankhauser, G., J. A. Vernon, W. H. Frank, and W. V. Slack. 1955. Effect of size and number of brain cells on learning in larvae of the salamander, *Triturus viridiscens. Science* 122:692-693.

Feldman, M. W., and L. L. Cavalli-Sforza. 1976. Cultural and biological evolutionary processes, selection for a trait under complex transmission. *Theor. Pop. Biol.* 9:238-259.

Fouts, R. S., and R. L. Rigby. 1977. Man-chimpanzee communication. In *How Animals Communicate*, ed. T. A. Sebeok, pp. 1034-1054. Bloomington: Indiana Univ. Press.

Gardner, R. A., and B. T. Gardner. 1969. Teaching sign language to a chimpanzee. *Science* 165:664-672.

Givnish, T. J., and G. J. Vermeij. 1976. Sizes and shapes of liane leaves. *Amer. Nat.* 110:743-778.

Goldschmidt, R. B. 1938. *Physiological Genetics.* New York: McGraw Hill.

Goodall, J. 1964. Tool using and aimed throwing in a community of free-living chimpanzees. *Nature* 201:1264-1266.

——. 1965. Chimpanzees of the Gombe Stream Reserve. In *Primate Behavior: Field Studies of Monkeys and Apes*, ed. I. DeVore, pp. 425-473. New York: Holt, Rinehart and Winston.

Gould, S. J. 1977. *Ontogeny and Phylogeny*. Cambridge, Mass.: Belknap Press of Harvard Univ. Press.

Greenough, W. 1975. Experimental modification of the developing brain. *Amer. Scientist* 63:37-46.

Griffin, D. R. 1976. The question of animal awareness. New York: Rockefeller Univ. Press.

Haldane, J.B.S. 1956. Time in biology. *Science Progress* 175:385-402.

Hamilton, W. D. 1964. The genetical theory of social behaviour, I, II, *J. Theor. Biol.* 7:1-52.

Healey, R. F., F. Cooke, and P. W. Colgan. 1979. Demographic consequences of snow goose brood rearing traditions. *J. Wildl. Manag.* (In press.)

Heglund, N. C., C. R. Taylor, and T. A. McMahon. 1974. Scaling stride frequency and gait to animal size. *Science* 186:1112-1113.

Hill, A. V. 1950. The dimensions of animals and their muscular dynamics. *Sci. Progress* 38:209-230.

Hinde, R. A., and J. Fisher. 1951. Further observations on the opening of milk bottles by birds. *Brit. Birds* 44:393-396.

Holt, A. B., D. B. Cheek, E. D. Mellits, and D. E. Hill. 1975. Brain size and the relation of the primate to the nonprimate. In *Fetal and Postnatal Cellular Growth: Hormones and Nutrition*, ed. D. B. Cheek, pp. 23-44. New York: John Wiley.

Hooker, T., and B. I. Hooker. 1969. Duetting. In *Bird Vocalizations: Their Relations to Current Problems in Biology and Psychology*, ed. R. A. Hinde, pp. 185-205. Cambridge: Cambridge Univ. Press.

Horn, H. S. 1968. The adaptive significance of colonial nesting in the Brewer's blackbird (*Euphagus cyanocephalus*). *Ecology* 49:682-694.

——. 1971. *The Adaptive Geometry of Trees*. Princeton: Princeton Univ. Press.

——. 1975. Forest Succession. *Sci. Amer.* 232 (5):90-98.

Hrdy, S. 1978. *The Langurs of Abu: Female and Male Strategies of Reproduction*. Cambridge, Mass.: Harvard Univ. Press.

Hubel, D. H. 1963. The visual cortex of the brain. *Sci. Amer.* 209 (5):54-62.

Jaynes, J. 1977. *The Origins of Consciousness in the Breakdown of the Bicameral Mind*. New York: Houghton Mifflin.

Jellis, R. 1977. *Bird Sounds and Their Meaning*. British Broadcasting Corp.

Jenkins, P. F. 1978. Cultural transmission of song patterns and dialect development in a free-living bird population. *Anim. Behav.* 25:50-78.

Jenni, D. A. 1974. Evolution of polyandry in birds. *Amer. Zool.* 14:129-144.

Jenni, D. A., and G. Collier. 1972. Polyandry in the American jacaña (*Jacaña spinosa*). *Auk* 89:743-765.

Jennings, H. S. 1905. *Behavior of the Lower Organisms*, reprint ed. Bloomington: Indiana Univ. Press, 1962.

Jerison, H. J. 1973. *Evolution of the Brain and Intelligence*. New York: Academic Press.

——. 1976. Paleoneurology and the evolution of mind. *Sci. Amer.* 234(1):90-101

Kalmus, H. 1955. The discrimination by the nose of the dog of individual human odours and in particular of the odours of twins. *Brit. J. Anim. Behav.* 3:25-31

Kandel, E. R., M. Brunelli, J. Byrne, and V. Castellucci. 1975. A common presynaptic locus for the synaptic changes underlying short term habitation and sensitization in the gill-withdrawal reflex in *Aplysia. Cold Spring Harbor Symp. Quant. Biol.* 40:465-482.

Kerr, W. E. 1950. Genetic development of castes in the genus *Melipona. Genetics* 35:143-152.

Kettlewell, H.B.D. 1955. Recognition of appropriate backgrounds by the pale and black phases of lepidoptera. *Nature* 175:943-944.

——. 1973. *The Evolution of Melanism*. Oxford: Clarendon Press.

Kimura, M., and T. Ohta. 1971 *Theoretical Aspects of Population Genetics*. Princeton: Princeton Univ. Press.

King, A. P., and M. J. West. 1977. Species identification in the North American cowbird: Appropriate responses to abnormal song. *Science* 195:1002-1004.

King, M-C., and A. C. Wilson. 1975. Evolution at two levels in humans and chimpanzees. *Science* 188:107-116.

Konishi, M. 1965. The role of auditory feedback in the control of vocalizations in the white-crowned sparrow. *Z. Tierpsychol.* 22:770-783.

Krebs, J., and R. M. May. 1976. Social insects and the evolution of altruism. *Nature* 260:9-10.

Kummer, H. 1968. *Social Organization of Hamadryas Bahoons: a Field Study.* Chicago: Chicago Univ. Press.

Lehrman, D. 1964. The reproductive behavior of ring doves. *Sci. Amer.* 211 (5):48-54.

Leigh, E. G. 1972. The golden section and spiral leaf arrangement. In *Growth by Intussusception*, ed. E. S. Deevy, Jr., pp. 161-176. Hamden, Conn.: Archon Books.

Light, S. F. 1942-1943. The determination of castes of social insects. *Quart. Rev. Biol.* 17:312-326, 18:46-63.

Lindaur, M. 1960. Time compensated sun orientation in bees. *Cold Spring Harbor Symp. Quant. Biol.* 25:371-377.

——. 1961. *Communication Among Social Bees.* Cambridge, Mass.: Harvard Univ. Press.

Lorenz, K. 1977. *Behind the Mirror. A Search for a Natural History of Human Knowledge.* London: Methuen.

Lüscher, M. 1961. Social control of polymorphism in termites. *Symp. Roy. Entomol. Soc. London* 1:57-67.

MacArthur, R. H. 1968. The theory of the niche. In *Population Biology and Evolution*, ed. R. C. Lewontin, pp. 159-176. Syracuse: Syracuse Univ. Press.

——. 1972. *Geographical Ecology.* New York: Harper and Row.

McClintock, M. 1971. Menstrual synchrony and suppression. *Nature* 229:244-245.

McNab, R. M., and D. E. Koshland. 1972. The gradient sensing mechanism in bacterial chemotaxis. *Proc. Nat. Acad. Sci. U.S.* 69:2509-2512.

McPhail, J. D., 1969. Predation and the evolution of stickleback (*Gasterosteus*). *J. Fish. Res. Bd. Can.* 26:3183-3208.

Marler, P. 1976. Organization, communication and graded signals: the chimpanzee and the gorilla. In *Growing Points in Ethology*, eds. P.P.G. Bateson and R. A. Hinde, pp. 239-280. Cambridge: Cambridge Univ. Press.

Marler, P., and M. Tamura. 1964. Culturally transmitted patterns of vocal behavior in sparrows. *Science* 146:1483-1486.

Mason, W. A. 1965. The social development of monkeys and apes. In *Primate Behavior: Field Studies of Monkeys and Apes*, ed. I. DeVore, pp. 514-543. New York: Holt, Rinehart and Winston.

May, R. M. 1977. Population genetics and cultural inheritance. *Nature* 268:11-13.

——. 1979. When to be incestuous. *Nature* 279:192.

Maynard Smith, J. 1975. *The Theory of Evolution*, 3rd ed. London: Penguin.

Maynard Smith, J. 1976. Evolution and the theory of games. *Amer. Sci.* 64:41-45.

——. 1977. Parental investment: A prospective analysis. *Anim. Behav.* 25:1-9.

——. 1978. *The Evolution of Sex.* Cambridge: Cambridge Univ. Press.

Mayr, E. 1963. *Populations, Species and Evolution.* Cambridge, Mass.: Belknap Press of Harvard Univ. Press.

——. 1974. Behavior programs and evolutionary strategies. *Amer. Sci.* 62:650-659.

Mech, L. D. 1970. *The Wolf: The Ecology and Behavior of an Endangered Species.* New York: Natural History Press.

Menzel, R., J. Erber, and T. Masuhr. 1975. Learning and memory in the honey bee. In *Experimental Analyses of Insect Behavior,* ed. L. Barton-Browne, pp. 195-217. New York: Springer-Verlag.

Mitchison, G. J. 1977. Phyllotaxis and the Fibonacci series: An explanation is offered for the characteristic spiral leaf arrangement found in many plants. *Science* 196:270-275.

Murie, A. 1944. *The Wolves of Mount McKinley.* Fauna of the National Parks of the United States, Fauna Series no. 5. Washington, D.C.: U S. Dept. of the Interior.

Naef, A. 1926. Über die Urformen der Anthropomorphen und die Stamesgeschichte des Menschenschädels. *Naturwiss.* 14:445-452.

Napier, J. R., and P. H. Napier. 1967. *A Handbook of Living Primates.* New York: Academic Press.

Norton-Griffiths, M. N. 1967. Some ecological aspects of the feeding behavior of the oystercatcher *Haematopus ostralegus* on the edible mussel *Mytilus edulis. Ibis* 109:412-424.

——. 1969. The organization, control and development of parental feeding in the oystercatcher (*Haematopus ostralegus*). *Behaviour* 34:55-114.

Nottebohm, F. 1972. The origins of vocal learning. *Amer. Nat.* 106:116-140.

Nottebohm, F., and R. K. Selander. 1972. Vocal dialects and gene frequencies in the Chingolo sparrow (*Zonotrichia capensis*). *Condor* 74:137-143.

Orians, G. H. 1969. On the evolution of mating systems in birds and mammals. *Amer. Natur.* 103:589-603.

Oster, G. F., and E. O. Wilson. 1978. *Caste and Ecology in the Social Insects.* Princeton: Princeton Univ. Press.

Pilbeam, D., and S. J. Gould. 1974. Size and scaling in human evolution. *Science* 186:892-901.

Post, R. H. 1971. Possible cases of relaxed selection in civilized populations. *Humangenetik* 13:253-284.

Quinn, W. G., and Y. Dudai. 1976. Memory phases in *Drosophila. Nature* 262:576-577.

Radinsky, L. 1978. Evolution of brain size in carnivores and ungulates. *Amer. Nat.* 112:815-831.

Rensch, B. 1956. Increase of learning ability with increase of brain size. *Amer. Natur.* 90:81-95.

——. 1960. *Evolution Above the Species Level.* New York: Columbia Univ. Press.

Richerson, R. J. and R. Boyd. 1978. A dual inheritance model of human evolutionary process I: basis postulates and a simple model. *J. Social Biol. Struct.* 1:127-154.

Richter, C. 1942. Total self regulatory functions in animals and human beings. *Harvey Lectures Series* 38:63-103.

Rozin, P. 1976. The selection of foods by rats, humans, and other animals. *Advances in the Study of Behavior*, vol. 6, eds. J. S. Rosenblatt, R A. Hinde, E. Shaw, C. Beer, pp. 21-76. New York: Academic Press.

Sahlins, M. 1976. *The Use and Abuse of Biology: An Anthropological Critique of Sociobiology.* Ann Arbor: The Univ. of Michigan Press.

Schmalhausen, I. I. 1949. *Factors of Evolution.* Philadelphia: Blakiston.

Schneirla, T. C. 1953. Modifiability in insect behavior. In *Insect Physiology*, ed. K. D. Roeder, pp. 723-747. New York: John Wiley and Sons.

——. 1971. *Army Ants.* San Francisco: W. H. Freeman.

Sebeok, T. A. 1977. *How Animals Communicate.* Bloomington: Indiana Univ. Press.

Shepher, J. 1971. Mate selection among second-generation kibbutz adolescents and adults. *Arch. Sex. Behav.* 1:293-307.

Short, R. V. 1976. The evolution of human reproduction. *Proc. Roy. Soc. London. B.* 195:3-24.

Silver, R. 1978. The parental behavior of ring doves. *Amer. Scient.* 66:209-215.

Silverman, M., and M. Simon. 1974. Flagellar rotation and the mechanism of bacterial motility. *Nature* 249:73-74.

Simpson, G. G. 1973. The Baldwin effect. *Evolution* 6:342.

Smith, W. J. 1977. *The Behavior of Communicating.* Cambridge, Mass.: Harvard Univ. Press.

Sturtevant, A. H. 1929. The *claret* mutant type of *Drosophila simulans*: A study of chromosomal elimination and cell lineage. *Z. Wiss. Zool.* 135:323-356.

Thompson, D'Arcy W. 1917. *On Growth and Form.* Cambridge: Cambridge Univ. Press.

Thorpe, W. H. 1972. *Duetting and Antiphonal Song in Birds; Its Extent and Significance.* In collaboration with J. Hall-Craggs, B. and T. Hooker and R. Hutchinson. Leiden: E. J. Brill.

Tinbergen, N. 1951. *The Study of Instinct.* Oxford: Clarendon Press of Oxford Univ. Press.

Trivers, R. L. 1971. The evolution of reciprocal altruism. *Quart. Rev. Biol.* 46:35-57.

Trivers, R. L. 1972. Parental investment and sexual selection. In *Sexual Selection and the Descent of Man*, ed. B. Campbell, pp. 136-179. Chicago: Aldine Pub. Co.

Trivers, R. L., and H. Hare. 1976. Haplodiploidy and the evolution of the social insects. *Science* 191:249-263.

von Frisch, K. 1967. *The Dance Language and Orientation of Bees*. Cambridge, Mass.: Belknap Press of Harvard Univ. Press.

Waddington, C. H. 1957. *The Strategy of the Genes: A Discussion of Some Aspects of Theoretical Biology*. London: George Allen and Unwin.

Washburn, S. L., and D. A. Hamburg. 1965. The implications of primate research. In *Primate Behavior: Field Studies of Monkeys and Apes*, ed. I. DeVore, pp. 607-622. New York: Holt, Rinehart and Winston.

Wecker, S. C. 1964. Habitat selection. *Sci. Amer.* 211(4):109-116.

Wells, M. J. 1962. *Brain and Behavior in Cephalopods*. Stanford, Calif.: Stanford Univ. Press.

West-Eberhardt, M. J. 1978. Temporary queens in Metapolybia wasps: nonreproductive helpers without altruism. *Science* 200:441-443.

Wicklund, C. 1972. Pupal coloration in *Papilio machaon* in response to the wavelength of light. *Naturwissenschaften* 59:219.

——. 1975. Pupal colour polymorphism in *Papilio machaon* L. and the survival in the field of cryptic versus non-cryptic pupae. *Trans. Roy. Ent. Soc. London* 127:73-84.

Williams, G. C. 1966. *Adaptation and Natural Selection; A Critique of Some Current Evolutionary Thought.* Princeton: Princeton Univ. Press.

——. 1975. *Sex and Evolution*. Princeton: Princeton Univ. Press.

Wilson, E. O. 1971. *The Insect Societies*. Cambridge, Mass.: Belknap Press of Harvard Univ. Press.

——. 1975. *Sociobiology: the New Synthesis*. Cambridge, Mass.: Belknap Press of Harvard Univ. Press.

——. 1976. The central problems of sociobiology. In *Theoretical Ecology*, ed. R. M. May, pp. 205-217. Philadelphia: W. B. Saunders Co.

——. 1978. *On Human Nature*. Cambridge, Mass.: Harvard Univ. Press.

Wireman, J. W., and M. Dworkin. 1975. Morphogenesis and developmental interactions in myxobacteria. *Science* 189:516-523.

Woolfenden, G. E. 1973. Nesting and survival in a population of Florida scrub jays. *Living Bird* 12:25-49.

——. 1975. Florida scrub jay helpers at the nest. *Auk* 92:1-15.

Woolfenden, G. E., and J. W. Fitzpatrick. 1978. The inheritance of territory in group-breeding birds. *BioScience* 28:104-108.

Wright, S. 1931. Evolution in Mendelian populations. *Genetics* 16:97-159.

Index